A First Course in Loop Quantum Gravity

Rodolfo Gambini and Jorge Pullin

OXFORD

UNIVERSITY PRESS

OXFORD

UNIVERSITY PRESS

Great Clarendon Street, Oxford OX2 6DP

Oxford University Press is a department of the University of Oxford.
It furthers the University's objective of excellence in research, scholarship,
and education by publishing worldwide in

Oxford New York

Auckland Cape Town Dar es Salaam Hong Kong Karachi
Kuala Lumpur Madrid Melbourne Mexico City Nairobi
New Delhi Shanghai Taipei Toronto

With offices in

Argentina Austria Brazil Chile Czech Republic France Greece
Guatemala Hungary Italy Japan Poland Portugal Singapore
South Korea Switzerland Thailand Turkey Ukraine Vietnam

Oxford is a registered trade mark of Oxford University Press
in the UK and in certain other countries

Published in the United States
by Oxford University Press Inc., New York

© R. Gambini and J. Pullin 2011

The moral rights of the authors have been asserted
Database right Oxford University Press (maker)

First published 2011

Reprinted 2012 (twice)

British Library Cataloguing in Publication Data

Data available

Library of Congress Cataloging in Publication Data

Data available

Typeset by SPI Publisher Services, Pondicherry, India
Printed and bound by
CPI Group (UK) Ltd, Croydon, CR0 4YY

ISBN 978-0-19-959075-9

3 5 7 9 10 8 6 4

Preface

Loop quantum gravity has emerged as a possible avenue towards the quantization of general relativity. Looking at the top 50 cited papers of all time of the arXiv.org:gr-qc preprint repository (which includes papers on gravity as a whole, not only quantum gravity), according to the SPIRES database in SLAC in its latest edition (2006) one finds that 13 papers are on loop quantum gravity. Although a smaller field than string theory, the other main approach to quantum gravity pursued today, the number of researchers working in loop quantum gravity is significant. At the moment both string theory and loop quantum gravity remain incomplete paradigms and as a consequence controversies about which is a more promising approach naturally arise. We will address some of these in this book.

Many people, including of course physics undergraduates, are interested in learning about loop quantum gravity. There are indeed excellent recent textbooks by Rovelli (2007) and Thiemann (2008) for those who want to pursue the topic in depth, roughly speaking at the level of an advanced US graduate school course. This type of treatment however, is largely inadequate for physics undergraduates and for others who want to get some minimal grasp of the subject in a relatively short period of time and without the depth expected for someone that is to do research in the field. An additional complication is that these "graduate-level" treatments of the subject have knowledge of general relativity as a prerequisite. This becomes a barrier for many readers. Although recent books by Hartle (2003) and Schutz (2009) make teaching general relativity at an undergraduate level possible, most undergraduate students have curricula that place general relativity at the very end of their careers (if their institution offers the course at all), which makes such courses unsuitable as prerequisites for a course in loop quantum gravity. Something that is not always appreciated is that undergraduates, particularly in the last semester of their career, are busy individuals. Most are taking several courses, perhaps conducting some undergraduate research work, preparing for the graduate record examinations (GRE), and applying to graduate schools, sometimes on top of holding a job. The time available to a specific course is very limited.

In these notes we will attempt to introduce loop quantum gravity without assuming previous knowledge of general relativity. The only background we will assume is knowledge of Maxwell's electromagnetism, a minimal knowledge of Lagrangian and Hamiltonian mechanics, and special relativity and quantum mechanics. This will inevitably imply we will be taking *many* shortcuts in the coverage of loop quantum gravity, but this will be the price we pay to introduce the topic in a way that is widely accessible to undergraduates at most physics programs in the US within the confines of a one-semester three-credit course. *Some* undergraduates and other readers may feel slightly disappointed in not getting more details and a more complete picture, but we believe the majority will welcome a book that is light and nimble as a good introduction to a subject that may otherwise appear intimidating. Some experts may feel we are short-changing readers by oversimplifying several issues for the sake of expediency. We will try to be careful to warn readers when we are doing so. Another goal we had in mind was to create a *short book*. Given that we are only introducing people to the topics, being deliberately superficial, and not attempting a full discussion, it is not worth trying to be exhaustive and discussing all issues in full detail. Long books tend to be intimidating to the reader and we think we will serve a larger audience with a compact book.

The organization of this book is as follows: in Chapter 1 we address the question of why one should quantize gravity. Chapter 2 will review Maxwell's electromagnetism and in particular its relativistic formulation. Chapter 3 will introduce some minimal elements of general relativity. Chapter 4 will deal with the Hamiltonian formulation of mechanics and field theories, including constraints. Chapter 5 will discuss Yang–Mills theories. Chapter 6 will cover quantum mechanics and some elements of quantum field theory. Chapter 7 introduces Ashtekar's new variables for general relativity. Chapter 8 develops the loop representation for general relativity. Chapter 9 presents quantum cosmology as an application. Chapter 10 discusses several miscellaneous applications including black hole entropy, the master constraint program and uniform discretizations, spin foams, possible experimental signatures, and the problem of time. The book ends with a chapter on the controversies surrounding loop quantum gravity.

Acknowledgements

We have benefited enormously from detailed comments on a draft version of the book by several people, who in many cases clearly spent a lot of time trying to help us: Fernando Barbero, Martin Bojowald, Steve Carlip, Jonathan Engle, Kristina Giesel, Gaurav Khanna, Kirill Krasnov, Daniele Oriti, Carlo Rovelli, Parampreet Singh, Madhavan Varadarajan, and Richard Woodard. To them we will always be grateful. This work was supported in part by the Gravitational Physics Program of the National Science Foundation, The Foundational Questions Institute (fqxi.org), the Horace Hearne Jr. Institute for Theoretical Physics, CCT-LSU and PDT (Uruguay).

A Gabriela y Martha

Contents

1
Why quantize gravity?

In our current understanding, there exist four fundamental interactions in nature: electromagnetism, weak interactions, strong interactions, and gravity. Everyone is familiar with electromagnetism. Weak interactions are involved in the decay of nuclei. Strong interactions keep nuclei together. The rules of quantum mechanics have been applied to electromagnetism, the weak and strong interactions. It is sort of natural to apply the rules of quantum mechanics to such interactions since they play key roles in the dynamics of atoms and nuclei and one knows that at such scales classical mechanics does not give correct predictions. The rules of quantum mechanics have not been applied to gravity in a satisfactory manner up to now. Loop quantum gravity is an attempt to do so, but it is an incomplete theory.

Before continuing we should clarify that from now on "gravity" is meant to be described not by the theory of Newton that one learns in elementary physics courses, but by Einstein's general theory of relativity. Numerous experimental tests of high accuracy agree with the predictions of general relativity (Will 2005). In such description, gravity is not really an "interaction" but rather a deformation of space-time. The latter is not flat and therefore objects do not follow naturally straight trajectories in space-time. This accounts for what one normally perceives as a gravitational "force," which in reality does not exist as such. In everyday parlance, we reinterpret the curved space-time around us as generating a gravitational force.

As we will discuss in this book, the fact that gravity is not a force but a deformation of space-time will make its quantization harder. The standard computational techniques of quantum field theories all assume one works on a given background space-time, even though in the end the goal is to create an S-matrix assuming that space-time is flat in the asymptotic past, future, and spatial infinity. But in gravity space-time is a field and therefore the object to be quantized without the presence of any background space-time. The lack of a background structure technically translates itself into the theory being naturally invariant under diffeomorphisms[1] of the space-time points, since there is nothing to distinguish one point from another. There is little experience in applying quantum field theory techniques to diffeomorphism invariant theories, except for certain topological theories with no local degrees of freedom. Moreover, gravity is not a very important force in the microscopic realm, where one expects quantum effects to be dominant. To understand this, consider the ratio of gravitational to electromagnetic forces between, say, a proton and an electron; gravity is about 10^{-40} times weaker. This is the root cause of why even today we do not have a single experiment that clearly requires quantum gravity for its explanation. It is perhaps the first time in the history of physics that one is trying to build a theory without experimental guidance. If gravity is harder to quantize, and in domains where it really is important quantum effects are expected to be small, why bother quantizing it at all? Could we not keep it as a classical interaction? This is not a moot question: attempts to quantize gravity have been made since the 1930s. If in nearly 80 years we have not succeeded, why keep trying?

To begin with, even though there are no calls from experimentally accessible situations where one needs to quantize gravity for their description, there are many physical processes one can imagine that require a quantum theory of gravity for their description. A simple example would be to study the collision of two particles at energies so high that gravity becomes relevant. Another example we will discuss later in the book would be to consider a black hole that evaporates via Hawking radiation until its mass is comparable to Planck's mass (10^{-5}g). Or another point we will cover later in the book: what happened to the universe close to the Big Bang? There is also the issue of conceptual clarity and unity in physics that suggests that quantizing three of the four fundamental interactions while keeping gravity classical is unsatisfactory. It is to be noted that the point of view of unification of theories (or more precisely unification of frameworks underlying theories) has been highly successful in history. For instance, putting Newton's mechanics and Maxwell's electromagnetism on the same footing led to special relativity. Incorporating gravity into

---

[1]A diffeomorphism is a map that assigns to each point of the manifold another point and that it is differentiable. It essentially "moves the points of the manifold around."

this framework led to general relativity. Incorporating special relativity into quantum mechanics led to quantum field theory. Similarly, unifying electromagnetism and the weak interactions led to the first satisfactory theory of the latter. In all instances putting theories on the same footing has led to the prediction of new physics, some of which have had dramatic implications. For instance, incorporating gravity, an apparently weak force, into special relativity leads to the notion of black holes and the Big Bang. Quantum field theory led to the notion of particles and antiparticles being created all the time in the vacuum.

Moreover, since we do not have a complete theory of quantum gravity it is hard to argue precisely that there are no experimental consequences of such a potential theory. There certainly are unexplained phenomena related to gravity out there, for instance those associated with dark energy and dark matter in cosmology, that—some conjecture—may eventually require a modification of the theory of gravity. That quantum gravity is in any way responsible for those effects remains to be seen.

But if we ignore theoretical considerations, is there a practical need to quantize gravity? As we argued before, there are no outstanding experiments that we know of that require a theory of quantum gravity for their explanation. There is no conclusive answer to this point, but it can be argued that it will be hard to have a consistent theory of classical gravity interacting with quantum fields. One quickly runs into problems with the uncertainty principle. Eppley and Hannah (1977) and Page and Geilker (1981) have devised thought experiments that illustrate this point (see however Mattingly (2006) for criticisms). For instance, consider a quantum object in a state with very small uncertainty in momentum and therefore a large uncertainty in position. We measure its position with great accuracy using a very sharp package of (classical) gravitational waves. That will require a superposition involving waves of very high frequency, which, however, in classical gravity can have arbitrarily small momentum. Through the measurement of the position with great accuracy, the uncertainty in momentum in the quantum system has suddenly become very large. We have therefore potentially produced a large change in the momentum of the total system, suggesting conservation of momentum may be violated.

Such experiments are not conclusive proof, since they cannot be carried out in practice (the above one has the problem that gravitational waves are very difficult to generate and control due to the weakness of the gravitational interaction). Carlip (2008) has argued that to consistently couple a classical gravitational field to a quantum system will require non-linear modifications of quantum mechanics that could be experimentally tested in some future.

In addition to this the two main paradigms of physics, general relativity and quantum field theories, have problems of their own. In general relativity powerful mathematical theorems proved in the 1960s and 1970s indicate that under generic conditions space-times become singular. Examples of such singularities are the Big Bang we believe to be present at the origin of the universe and the singularities that arise inside black holes. A singularity generically is associated with a divergence in quantities that indicates the theory has been pushed beyond its realm of applicability. Near such singularities one usually encounters energy densities that are not compatible with a completely classical treatment anymore. The expectation is therefore that a theory unifying quantum field theory with general relativity could offer a new perspective, and perhaps eliminate the singularities altogether.

Similarly, the quantum theory of fields has the problem that many quantum operators are in the mathematical sense not functions but distributions, like the Dirac delta. When one studies interactions one has to consider products of these operators and such products are usually not well defined. Some of the divergences in quantum field theories can be eliminated, redefining the coupling constants through a procedure known as renormalization, and one can use the theories to formulate physical predictions. In spite of this, as freestanding mathematical theories quantum field theories are usually poorly defined, and have to be treated using perturbative series that are not really convergent (they are what is technically known as asymptotic expansions). Singularities in quantum field theory arise due to the distributional nature of the fields and operators. It is clear that if one changes the underlying picture of space and time the distributional nature of fields and operators may change. This may open the possibility of eliminating the singularities of field theory.

Summarizing the last two points: the current physical paradigms for gravity and for field theories are incomplete and include singularities. There appears the distinct possibility that by unifying these paradigms, singularities could be eliminated. We will take the point of view that aesthetics, the previously mentioned thought experiments, and the attractive possibility that unification may cure the problems of the standalone paradigms of gravity and quantum field theory, are enough motivation to say that gravity needs to be quantized.

The previous discussion also highlights some of the open problems of the field that people are attempting to address in contemporary research. Is there a singularity at the Big Bang or did our current universe evolve from a previous universe? If so, are there remnants of information coming from the previous universe? Does such a potential modification of the Big Bang influence the rest of the cosmological evolution, in particular how inflation developed, nucleosynthesis, and the formation of structure in the universe? What happens inside a black hole when curvatures become large? Does one again travel into another space-time region? We now know that black holes radiate like black bodies and could eventually evaporate. How is such evaporation described in detail? Since the final product of the evaporation is purely thermal radiation with no distinctive features apart from its temperature, what happened to the information included in the matter that formed the black holes? Are there any phenomenological consequences of quantizing gravity that we could observe? We will touch upon all these topics in the applications chapter towards the end of the book.

Let us now turn to a bit of history of the field. We will not attempt a detailed history here, just give some minimal background. A good concise treatment of the history of quantum gravity is in the article by Rovelli (2002). Although one already encounters mention of the quantization of gravity in papers by Einstein in 1916 and Rosenfeld and Bronstein then wrote the first detailed papers on the subject in the 1930s, significant attempts to quantize gravity started only in the early 1960s. Three different approaches emerged. One approach was canonical quantization, which we will largely follow in this book, since it is the one that resembles the elementary treatments of quantum mechanics of undergraduate textbooks that students may be familiar with. In this approach one has to separate space-time into space and time and as we will see, this will add complications. It took quite a while to understand how to work out the Hamiltonian formulation of theories like the one that describes gravity. We will get a flavor in this book of why it took some effort. Another approach was to study the theory perturbatively, by assuming that space-time is flat plus small perturbations. Such perturbative approaches worked well for electromagnetism and the weak and strong interactions (in the latter case in certain particular regimes). In gravity this approach ran into trouble. In electromagnetism and the weak and strong interactions one can formulate the theory in terms of a coupling constant that is dimensionless (in electromagnetism, for instance, it is the fine structure constant). In gravity one cannot do that. Having a coupling constant with dimensions implies that if one makes expansions in powers of the coupling constant, as one does in perturbation theory, extra powers of momentum have to be introduced at each order to keep the expression dimensionally uniform. The extra powers of momentum make the integrals arising in the interaction terms divergent. One can correct such divergences by modifying the action, but it requires an infinite number of modifications to cure all divergences. To have to specify an infinite number of terms by hand implies that the theory does not have predictive power. This problem is known as non-renormalizability. Stelle (1977) showed that one could cure the problem by adding some higher order terms to the action, but the resulting theory of gravity has unphysical properties. A good review of the perturbative approach is that of Woodard (2009). We will present a highly simplified discussion in this book. The third approach that was tried is the one known as Feynman path integral. Such an approach requires summing probability amplitudes over all classical trajectories, which in the case of gravity requires summing over all possible space-times. This has proved formidably difficult to do. Remarkably, loop quantum gravity techniques are helping define the path integral in a rigorous way, in an approach called *spin foams*. We will cover spin foams only briefly in this book.

At the same time as these approaches were encountering difficulties, a parallel line of thought was being pursued, namely that of unification of the elementary interactions into a single theory. A motivation for this comes from the weak interactions: it turns out that one cannot quantize the weak interactions by themselves, but only when they are integrated into a theory that unifies them with electromagnetism. Could the situation be similar in gravity? Could it be that integrating gravity with the other interactions into a single theory would help with its quantization? This point of view tends to be favored by most physicists with backgrounds in particle physics. Over the years it has led to a series of theories that attempt to unify gravity with the other interactions and at the same time provide a theory of quantum gravity. The various approaches included the Kaluza–Klein theories, supergravity, and lately, string theory and M-theory. We will not attempt a discussion of these approaches here. A textbook for undergraduates on string theory is available by Zwiebach (2009).

In addition to the approaches described above, there are other ideas that are pursued by smaller groups of researchers. These include causal dynamical triangulations, causal sets, matrix models, Regge calculus, twistors, noncommutatiave geometries, and the asymptotic safety scenario. We will not discuss them in this book. A good introductory overview is presented in the book by Smolin (2002).

In the mid 1980s, Ashtekar noted that one could rewrite the equations of gravity in terms of variables that made the theory resemble the theories of particle physics. This raised hopes that techniques from particle physics could be imported to the quantization of gravity. The resulting approach to quantizing gravity is called "loop quantum gravity" and is the one we will cover in this book. It is an attempt to understand the quantization of gravity by itself, without the need to unify it with other interactions.

Currently, both string theory and loop quantum gravity are incomplete theories. Some people view them as competing theories and imply that if one ends up being correct the other will not be. Our point of view is more conservative. It could end up being that string theory and loop quantum gravity both provide quantum theories of gravity cast in different language and highlighting different aspects of the problem in more natural ways in each approach. At the moment it is still unclear if this is the case.

2

Special relativity
and electromagnetism

Newton's laws of mechanics take their simplest form in certain reference frames called inertial frames. One cannot distinguish a preferred inertial frame, they are all equivalent, and in all of them Newton's laws take the same form. This is the principle of Galilean relativity. However, when one considers electromagnetism as formulated by Maxwell there do appear to exist preferred frames of reference. This caught the attention of Einstein, who found the situation unsatisfactory. He was particularly troubled that in order to describe a magnet moving close to a wire, or a wire moving close to a magnet, one needed to use two different physical laws even though the end result, the production of a current in the wire, was exactly the same in both cases. This was only one of the many apparent conflicts that arose when trying to understand electromagnetic behavior in mechanical terms.

Einstein's observation was that all physical laws should be subject to the principle of relativity, they should take the same form in all inertial frames. In particular, if Maxwell's equations take the same form in all inertial frames, the speed of light should take the same value in all inertial frames. This last observation indicates that inertial frames cannot be related by Galilean transformations, since the latter do not keep the speed of light constant. If one accepts this point of view, Galilean transformations have to be abandoned along with the concomitant hypotheses about space and time that were used to build Newton's theory. In their place one needs to introduce new transformation laws to relate events viewed from different inertial reference frames. Such transformations are the Lorentz transformations. Galilean relativity was built on daily life observations that seemed rooted in common sense, but in reality were only observations made in a relatively narrow range of relative speeds. It turns out that Galilean relativity is only a slow speed approximation of a more fundamental (and more importantly, physically correct) relativity principle. The latter has many implications involving our ideas of space, time, and simultaneity. In this chapter we will explore some of these ideas and in particular introduce the mathematical notation for it that we will later use to describe general relativity.

2.1 Space and space-time

Let us start with some elementary vector notation in ordinary three-dimensional space. We start by setting up Cartesian coordinates in space x^i with $i = 1, 2, 3$. The distance between two points in space is given by Δs with,

$$\Delta s^2 = \left(\Delta x^1\right)^2 + \left(\Delta x^2\right)^2 + \left(\Delta x^3\right)^2 = \sum_{i=1}^{3} \left(\Delta x^i\right)^2 \tag{2.1}$$

with Δx^i the separation of the two points along the i-th coordinate. One can choose a new set of Cartesian axes, like those shown in Fig. 2.1 (for simplicity we draw a two-dimensional example), and the distance between two points remains unchanged,

$$\Delta s^2 = \sum_{i=1}^{3} \left(\Delta x^{i'}\right)^2 \tag{2.2}$$

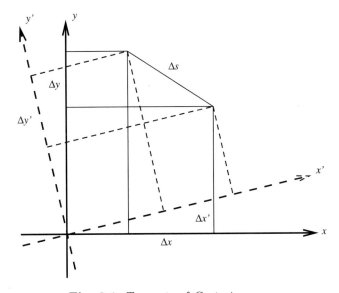

Fig. 2.1 *Two sets of Cartesian axes.*

where we have denoted by $x^{i'}$ the new set of axis. This notation of using a prime in the index rather than in the name of the coordinate will prove useful later on. If the rotation between the axes shown in the figure is given by an angle θ then one can relate the values of x^i and $x^{i'}$ by,

$$x^{i'} = \Lambda^{i'}{}_i x^i \equiv \sum_{i=1}^{3} \Lambda^{i'}{}_i x^i, \tag{2.3}$$

and also

$$\Delta x^{i'} = \Lambda^{i'}{}_i \Delta x^i, \tag{2.4}$$

where we have introduced the *Einstein summation convention,* in which any index that is repeated is summed over from one to three. At the moment the use of subscripts or superscripts in a given quantity makes no difference, but it will later on. The matrix Λ is given by

$$\Lambda^{i'}{}_i = \begin{pmatrix} \cos\theta & \sin\theta & 0 \\ -\sin\theta & \cos\theta & 0 \\ 0 & 0 & 1 \end{pmatrix}. \tag{2.5}$$

A vector is a collection of three numbers A^i that transform under coordinate transformations like the coordinates themselves, that is,

$$A^{i'} = \Lambda^{i'}{}_i A^i. \tag{2.6}$$

You are probably familiar with all this, we are just fixing notation here. Something you are perhaps not so familiar with is with the notion of *tensor.* A tensor is a multi-index generalization of a vector. The key idea is that each index in a tensor transforms as if it were sitting on a vector, without any regard to what happens to the other indices. For instance, if we consider a tensor with two indices S^{ij}, it will transform as,

$$S^{i'j'} = \Lambda^{i'}{}_i \Lambda^{j'}{}_j S^{ij} \tag{2.7}$$

where again we are assuming two summations, one on i and one on j.

In ordinary Newtonian physics, if one uses Cartesian coordinates, these are all the vector transformations one needs. In particular the time variable t remains unchanged. In special relativity the situation is different. Special relativity refers to space-time and involves transformations that mix space and time together. Remarkably, we will see that the language for coordinate transformations that we have described up to now applies almost unchanged to special relativity.

Let us now consider a space-time, that is, a four-dimensional space with coordinates x^μ with $\mu = 0, 1, 2, 3$, with the $1, 2, 3$ components coinciding with x^i, and the zeroth component will be given by ct where c is the speed of light. The reason we need c is to have the same units in all the components. In theoretical physics it is common to choose units in which $c = 1$. In this text we will choose this convention unless we want to emphasize the role of the speed of light, in which case we will state it explicitly. A choice of $c = 1$ requires measuring time in units of distance. A "point" in space-time is an "event," it is an assignment of a point in space and an instant in time.

Up to now there is nothing special being done from the point of view of physics. We could have set up a similar notation in Newtonian mechanics, bundling space and time into four-dimensional vectors. But it would not have been very useful. Since in Newtonian mechanics time is unchanging, we would only be doing transformations that involve the components $1, 2, 3$ of the four-vectors and leaving the zeroth unchanged. There would be nothing gained.

For something physically new to happen we need to introduce the idea of "distance" in space-time that is physically useful in special relativity. Such "distance" between two events is given by,

$$\Delta s^2 = -(c\Delta t)^2 + (\Delta x^1)^2 + (\Delta x^2)^2 + (\Delta x^3)^2. \tag{2.8}$$

Notice that this is not a true distance in that it can be positive, negative, or even zero (even for points that do not coincide). This distance is kept invariant under the *Lorentz transformations*, that is[1], $x^{\mu'} = \Lambda^{\mu'}{}_\mu x^\mu$,

$$\Delta s^2 = -(c\Delta t')^2 + (\Delta x^{1'})^2 + (\Delta x^{2'})^2 + (\Delta x^{3'})^2. \tag{2.9}$$

Lorentz transformations come in various kinds. First of all, ordinary spatial rotations are included, for instance the θ rotation we considered in Fig. 2.1 yields the following Lorentz transformation,

---

[1]When we are considering space-time quantities repeated indices are summed from 0 to 3.

$$\Lambda^{\mu'}{}_{\mu} = \begin{pmatrix} 1 & 0 & 0 & 0 \\ 0 & \cos\theta & \sin\theta & 0 \\ 0 & -\sin\theta & \cos\theta & 0 \\ 0 & 0 & 0 & 1 \end{pmatrix}. \tag{2.10}$$

We also have *boosts* which mean that we are changing to a new frame of reference that is moving with respect to the original one. For instance if we change to a coordinate system that is moving along the x^1 direction with velocity v with respect to the original one, we have,

$$\Lambda^{\mu'}{}_{\mu} = \begin{pmatrix} \cosh\phi & -\sinh\phi & 0 & 0 \\ -\sinh\phi & \cosh\phi & 0 & 0 \\ 0 & 0 & 1 & 0 \\ 0 & 0 & 0 & 1 \end{pmatrix}, \tag{2.11}$$

with the parameter ϕ (which ranges from $-\infty$ to ∞) given by $\phi = \tanh^{-1}(v)$ (remember we are setting $c = 1$). If you explicitly work out the transformation $x^{\mu'} = \Lambda^{\mu'}{}_{\mu} x^{\mu}$ with the above matrix you will find that

$$t' = t\cosh\phi - x\sinh\phi, \tag{2.12}$$

$$x' = -t\sinh\phi + x\cosh\phi. \tag{2.13}$$

These expressions may appear unfamiliar, but a bit of algebra and trig identities leads to the familiar expressions,

$$t' = \gamma(t - vx), \tag{2.14}$$

$$x' = \gamma(x - vt), \tag{2.15}$$

where $\gamma = 1/\sqrt{1 - v^2}$. If the velocity v is small compared to the speed of light, which in our units means $v \ll 1$, then the above expressions can be approximated by $t' = t$ and $x' = x - vt$, that is, a Galilean transformation (to see this it is good to restore the speed of light in the equations which read $t' = \gamma(t - vx/c^2)$, the second is unchanged and $\gamma = \sqrt{1 - v^2/c^2}$ and $v \ll c$).

So, just like ordinary rotations keep spatial distances unchanged, Lorentz transformations keep unchanged the space-time distance we introduced in (2.8). Lorentz transformations mix space and time and it is therefore natural to think of space and time as a four-dimensional single entity. That minus sign that appears in the space-time distance, however, has rather significant implications. To see some of those, it is very useful

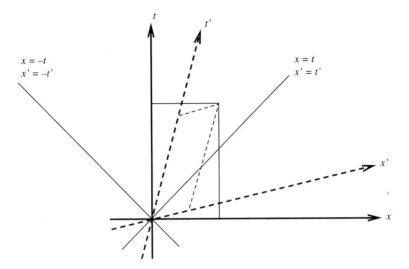

Fig. 2.2 *Effect of a Lorentz transformation in space-time.*

to consider some *space-time diagrams*. Let us suppress two of the three
spatial dimensions for simplicity and consider the Lorentz transformation,
specifically the boost in the x direction, graphically. In Fig. 2.2 we see
significant differences with respect to Fig. 2.1. This transformation is not
a rotation, but rather the axes seem to "close in" symmetrically around
the line $x = t$, which in turn remains invariant (as does the line $x = -t$),
that is it coincides with $x' = t'$. Notice that the line $x = \pm t$ corresponds
to the trajectory of an object moving (rightwards/leftwards) at the speed
of light in space-time. We know that the speed of light is invariant in
special relativity and the invariance of the $x = \pm t$ lines reflects that fact.
Remember that we are suppressing two spatial directions, if not, the $x = t$
line would correspond to a *light cone* $t^2 = (x^1)^2 + (x^2)^2 + (x^3)^2$. Since
the equation is quadratic, there is a *future light cone* and a *past light
cone* with respect to the origin of the diagram (cone pointing upwards or
downwards). Since the origin of the diagram was chosen arbitrarily, this
situation can apply to any point of space-time. That is, at every point
in space-time there is a future and past light cone. All points within
the future or past light cones of a given point are said to be *time-like
separated* with respect to the point in question. The space-time distance
between those points is negative. Points outside the light cone are *space-
like separated* and their distance to the origin is positive, like an ordinary
spatial distance. Finally, points along the light cone are *light-like* or *null*

separated and their distance to the origin is zero! Notice that in this picture the notion of simultaneity is not absolute. Consider the frame t, x. There, points at $t = 0$ are all simultaneous, they correspond to the horizontal line at the origin. However, if we consider the same line but in the t', x' frame, we see that it does not correspond to a single value of t', that is, those events are not simultaneous in that reference frame.

The non-universality of simultaneity can be very confusing to deal with at first. This leads to many "paradoxes." The simplest way to resolve such paradoxes consists of drawing a space-time diagram. Then it usually becomes very clear which preconception from spatial intuition have gone wrong. For instance a moving object's length appears to contract. This is known as "Lorentz contraction." Does that mean that one could trap the object inside a box into which it would not fit at rest just by having it move? This is the core of the "pole in the barn" paradox, which we leave as one of the exercises at the end of the chapter.

Some basics about space-time diagrams: as we mentioned, light rays move along the light cone, in the diagrams where one suppresses two dimensions it moves along the lines at 45 degrees $t = \pm x$ (with $c = 1$). A particle at rest moves upwards along a vertical line. All moving particles move at less than the speed of light, that is they describe trajectories steeper than 45 degrees with respect to the horizontal. Events that occur in the future light cone of a given event are *causally connected* to it, that is, they can be the consequence of said event. Events outside the light cone of an event cannot be caused by it, it would require communication faster than light for that to happen. Uniform motion in space-time corresponds to straight lines, the slope giving the velocity. Accelerated trajectories all eventually asymptote to $t = \pm x$ since nothing can move faster than light, even accelerating objects. A point to be noted is that when one goes to a moving frame, in addition to the "closing in" of the axis we mentioned above, there is a rescaling of the axis, which can be significant for speeds close to the speed of light. This needs to be taken into account when drawing space-time diagrams to clarify paradoxes.

2.2   Relativistic mechanics

Just like in ordinary space, one defines vectors in space-time as a collection of four numbers A^μ that transform like the coordinates, that is $A^{\mu'} = \Lambda^{\mu'}{}_\mu A^\mu$. As an example of a four-vector, consider a trajectory in space-time, that is, a curve $x^\mu(\lambda)$ where λ is a parameter. If we construct the tangent to the space-time curve associated with the trajectory $U^\mu \equiv dx^\mu/d\lambda$, one can quickly see that it transforms as the coordinates, since the transformation matrix Λ does not depend on λ.

A useful operation to have among vectors is that of a scalar product. We know that the scalar product of an ordinary vector with itself gives the length of the vector. In ordinary space we could define the scalar product in the following fashion, for instance, starting from the notion of distance,

$$\Delta s^2 = \delta_{ij} \Delta x^i \Delta x^j \tag{2.16}$$

where δ_{ij} is the identity matrix in three dimensions, that is it is one if $i = j$ and zero otherwise. So we would say that Δs^2 is the scalar product of Δx^i with itself. Similarly one can define a scalar product in four dimensions, except that the matrix $\eta_{\mu\nu}$ that is involved is not the identity matrix anymore,

$$\Delta s^2 = \eta_{\mu\nu} \Delta x^\mu \Delta x^\nu \tag{2.17}$$

but is given by $\eta_{\mu\nu} = \mathrm{diag}(-1, 1, 1, 1)$. This matrix is called the *Minkowski metric*. The word "metric" denotes that it helps us measure distances. But one can use the metric to define the scalar product of any vector with another or with itself. Notice that, just like the space-time distance, the scalar product of a four-vector with itself is not positive definite. If it is negative the vector is called *time-like* if positive *space-like* and if zero the vector is called *null* (there are variations in the literature as to which is defined as positive and which as negative).

The transformation matrix for vectors keeps the Minkowski metric invariant,

$$\eta_{\mu\nu} = \Lambda_\mu{}^{\mu'} \Lambda_\nu{}^{\nu'} \eta_{\mu'\nu'}, \tag{2.18}$$

and the notions of time-like, space-like, and null are coordinate invariant.

If we consider a curve in space-time we can measure the length along the curve,

$$\Delta l = \int \sqrt{\eta_{\mu\nu} \frac{dx^\mu}{d\lambda} \frac{dx^\nu}{d\lambda}} d\lambda. \tag{2.19}$$

One has to be careful since the argument of the square root could become negative. The above definition is correct as long as it is positive. That means the curve is space-like. If the curve is time-like, that is, each successive point is within the light cone of the previous one, then the tangent vector to the curve is time-like and the scalar product inside the square root is negative, so we define,

$$\Delta\tau = \int \sqrt{-\eta_{\mu\nu}\frac{dx^\mu}{d\lambda}\frac{dx^\nu}{d\lambda}}d\lambda. \qquad (2.20)$$

τ is called the *proper time*. The reason for this is that if we consider a space-time trajectory that does not move spatially, that is a particle at rest, then $\Delta\tau = \Delta t$, that is the time measured at rest with respect to the particle. Particle trajectories are always time-like or, if one considers massless particles, null. In that case the length of the trajectory vanishes.

We can use this concept to define a physically relevant vector, the four-velocity, $U^\mu \equiv dx^\mu/d\tau$. Since $d\tau^2 = \eta_{\mu\nu}dx^\mu dx^\nu$ we immediately see that $\eta_{\mu\nu}U^\mu U^\nu = -1$, that is the four-velocity is automatically normalized.

A related four-vector for a particle is the four-momentum, defined as $p^\mu = mU^\mu$. The quantity m is known as the *rest mass* of the particle. Such quantity is an invariant under Lorentz transformations and corresponds, in the rest frame, to the zeroth component of p^μ. Starting from the rest frame, where $p^\mu = (m, 0, 0, 0)$, we can go to a frame where the particle is moving with velocity v via a Lorentz transformation. Say we choose the boost in the x direction we have been discussing. In that case $p^{\mu'} = (\gamma m, v\gamma m, 0, 0)$, where as before $\gamma = 1/\sqrt{1 - v^2}$. Suppose we choose a small v, corresponding to a particle moving slowly with respect to the speed of light. If we Taylor expand $\gamma \sim 1 + v^2/2 + \dots$. So to leading order in velocities we have $p^0 = m + mv^2/2$ and $p^1 = mv$. We therefore see that for small velocities the spatial components of the four-momentum are just the ordinary momentum. What about the zeroth component? We identify in it the kinetic energy $mv^2/2$ but what about the term m? We can identify that term as the energy of a mass in its rest frame. This leads to the famous formula (restoring c) $E = mc^2$. That is, the zeroth component of the four-momentum is the energy of the particle including a contribution mc^2 from its rest mass and whatever contributions come from kinetic energy. So we see that this vector really captures the energy and momentum content of a particle.

To complete the formulation of relativistic mechanics, one introduces a four-dimensional version of Newton's law. For this one introduces the acceleration four-vector $a^\mu = dU^\mu/d\tau$, and one has that it is proportional to the four-force,

$$f^\mu = ma^\mu. \tag{2.21}$$

Notice that the four-force is not an arbitrary vector, since it has to be orthogonal to the four-velocity (this can be easily seen differentiating $\eta_{\mu\nu}U^\mu U^\nu = -1$). It therefore has only three independent components.

Before moving on to electromagnetism, we need to introduce another small item dealing with indices. In usual spatial vector algebra, one normally does not care where indices are placed. In space-time it is useful to make a distinction between subindices and superindices. Given a four-vector A^μ we define the same four-vector "with its index down" as $A_\mu \equiv \eta_{\mu\nu}A^\nu$. The two four-vectors are not the same, one of the components differs by a sign. One of the reasons this is useful is because one can take a vector with an index down, take another vector with the index up, and their (Lorentz invariant) scalar product is just obtained by summing the products of their components, that is $A_\mu B^\mu \equiv A^\mu B^\nu \eta_{\mu\nu}$. Similarly, given a vector with its index down we can "raise it" with $A^\mu = \eta^{\mu\nu}A_\nu$. The matrix $\eta^{\mu\nu} = \text{diag}(-1,1,1,1)$ coincides with $\eta_{\mu\nu}$. This sounds like a lot of fuss to deal with a simple minus sign. But later on in the book we will want to discuss space-times that are not flat. In this case $\eta_{\mu\nu}$ gets replaced by a more complicated, generically non-diagonal, non-constant matrix. In this case raising or lowering an index implies much more than just flipping a sign. If there is such a difference, why do we call the object with an index down the same name as the object with the index up? The reason is that the physical content of the object remains the same, it is just "rearranged" in a different form once the metric is given. Some physical quantities are naturally defined with their index down, some with their index up. For instance, the four-velocity is naturally defined with its index up, even in curved space-times. By "naturally" we mean that we do not need involve the metric of space-time in its definition with the index up, we do with the index down.

At this point it is convenient to mention some additional jargon that is used with tensors, the idea of symmetry and antisymmetry. A set of indices (either all lower or all upper but never mixed) are said to be symmetric if swapping any pair of them produces the same object. The simplest example is $\eta^{\mu\nu}$ which satisfies $\eta^{\mu\nu} = \eta^{\nu\mu}$ since it is a symmetric matrix. But the concept extends to many indices as well. A tensor, in contrast, is said to be antisymmetric in a set of (either upper or lower) indices, such that if one swaps a pair of them the tensor changes sign. In the next section we will see an example of antisymmetric tensor playing an important role.

2.3 Maxwell theory

Let us turn our attention to electromagnetism. Maxwell's theory of electromagnetism is a theory of two vector fields, called the electric field E^i and the (pseudo-vector) magnetic field B^i. Since we are interested in fundamental physics, we will only consider the theory in vacuum coupled to charges, that is we will not discuss material media. Maxwell's equations are,[2] written in traditional vector notation,

$$\vec{\nabla} \times \vec{B} - \frac{\partial \vec{E}}{\partial t} = 4\pi \vec{J}, \tag{2.22}$$

$$\vec{\nabla} \cdot \vec{E} = 4\pi\rho, \tag{2.23}$$

$$\vec{\nabla} \times \vec{E} + \frac{\partial \vec{B}}{\partial t} = 0 \tag{2.24}$$

$$\vec{\nabla} \cdot \vec{B} = 0, \tag{2.25}$$

where ρ is the density of charge and $\vec{J}$ is the current density. Let us attempt to rewrite these equations in space-time notation. To do this we recall the definition of a cross product of two vectors in index notation,

$$(A \times B)^i = \epsilon^{ijk} A_j B_k, \tag{2.26}$$

where the quantity ϵ^{ijk} is known as the Levi–Civita symbol and is equal to one if i, j, k are an even permutation of $1, 2, 3$ (that is $2, 3, 1$ or $3, 1, 2$), is zero if two indices are repeated and is -1 otherwise. We can write Maxwell's equations as,

---

[2]As is common in theoretical physics we choose natural units where $c = \epsilon_0 = \mu_0 = 1$.

$$\epsilon^{ijk}\partial_j B_k - \partial_0 E^i = 4\pi J^i, \tag{2.27}$$

$$\partial_i E^i = 4\pi J^0, \tag{2.28}$$

$$\epsilon^{ijk}\partial_j E_k + \partial_0 B^i = 0, \tag{2.29}$$

$$\partial_i B^i = 0. \tag{2.30}$$

Please recall that for spatial indices it makes no difference if we write them as subscripts or superscripts. The notation ∂_i means $\partial/\partial x^i$ and the notation ∂_0 means $\partial/\partial t$. Notice that we have replaced the charge density by J^0 so we are building a four-vector $J^\mu = (\rho, J^1, J^2, J^3)$.

We now proceed to define a *field tensor*,

$$F_{\mu\nu} = \begin{pmatrix} 0 & -E_1 & -E_2 & -E_3 \\ E_1 & 0 & B_3 & -B_2 \\ E_2 & -B_3 & 0 & B_1 \\ E_3 & B_2 & -B_1 & 0 \end{pmatrix} \tag{2.31}$$

and we note that the same object with its indices up is defined as $F^{\mu\nu} \equiv \eta^{\mu\rho}\eta^{\nu\sigma}F_{\rho\sigma}$. Both objects are antisymmetric $F_{\mu\nu} = -F_{\nu\mu}$. We can identify various components of the electric and magnetic fields, for example, $F^{0i} = E^i$, $F^{ij} = \epsilon^{ijk}B_k$ (it is a good exercise for you to check these). With this notation the first four Maxwell equations can be rewritten as,

$$\partial_j F^{ij} + \partial_0 F^{i0} = 4\pi J^i, \tag{2.32}$$

$$\partial_i F^{0i} = 4\pi J^0, \tag{2.33}$$

or, remarkably as,

$$\partial_\mu F^{\nu\mu} = 4\pi J^\nu. \tag{2.34}$$

If we now introduce the space-time Levi–Civita symbol $\epsilon^{\mu\nu\rho\sigma}$ defined again as $+1$ if μ, ν, ρ, σ are an even permutation of $0, 1, 2, 3$ (that is $1, 2, 3, 0$ or $2, 3, 0, 1$ or $3, 0, 1, 2$), zero if two indices repeat and -1 otherwise, one can rewrite the other four Maxwell equations as

$$\epsilon^{\sigma\mu\nu\lambda}\partial_\mu F_{\nu\lambda} = 0. \tag{2.35}$$

To see this, for instance if $\sigma = 0$, then $\mu, \nu,$ and λ have to be spatial for the Levi–Civita to be non-zero, so we can rewrite the equation as $\epsilon^{ijk}\partial_i F_{jk} = \epsilon^{ijk}\partial_i \epsilon_{jkm}B^m$ and then use the identity $\epsilon^{ijk}\epsilon_{jkm} = 2\delta^i_m$ to get $\partial_i B^i = 0$. A similar construction can be done for the three other values of σ. It is worth noticing that the Levi–Civita symbol is invariant under Lorentz transformations (up to an overall sign given by the determinant of $\Lambda_\mu{}^{\mu'}$).

Let us pause and see what has happened here. We have rewritten Maxwell's equations in space-time notation. The resulting equations have been expressed entirely in terms of vectors and tensors. That means we know how to transform these equations under Lorentz transformations and they will have the same form in all reference frames. So Maxwell's theory was secretly Lorentz invariant, we were just not writing it in a way that made that manifest. We now have. What do we mean by this? Well, suppose that you are given an electric and a magnetic field E^i, B^i and you are asked to transform them to a moving frame. Physically we know they are not invariant. For instance, consider a charge at rest. It has an electric field and no magnetic field. If we go to a frame where the charge is moving, there will be a magnetic and an electric field. Computing them in ordinary vector notation is not straightforward. However, in space-time notation it is easy, simply write the field tensor $F^{\mu\nu}$ and then apply the Lorentz transformation (remember, just treat every index as if the others didn't exist), that is $F^{\mu'\nu'} = \Lambda^{\mu'}{}_\mu \Lambda^{\nu'}{}_\nu F^{\mu\nu}$ and you will get the fields in the new frame. The transformation will mix together E^i and B^i as expected. We see here the space-time notation starting to bear fruits. And it reflects the spirit of Einstein we alluded to at the beginning of the chapter.

To conclude the discussion of Maxwell theory it is a good idea to introduce the notion of potentials. In electromagnetism one knows that there is an electrostatic potential ϕ and a vector potential $\vec{A}$ such that,

$$\vec{E} = -\vec{\nabla}\phi - \frac{\partial \vec{A}}{\partial t}, \tag{2.36}$$

$$\vec{B} = \vec{\nabla} \times \vec{A}. \tag{2.37}$$

To rewrite these equations in space-time notation, we begin by introducing a space-time vector potential,

$$A^\mu = (\phi, \vec{A}), \tag{2.38}$$

in terms of which we have,

$$F_{\mu\nu} = \partial_\mu A_\nu - \partial_\nu A_\mu. \tag{2.39}$$

Notice the simplicity of these expressions and how easy they are to remember. In particular you now know how to quickly transform potentials to moving reference frames. Notice, for instance, that if one has a charge at rest one can choose $\vec{A} = 0$, but if one goes to a frame where the charge is moving there will be a non-vanishing $\vec{A}$ as the components of A_μ get mixed by Maxwell's equations. If one assumes that the electromagnetic field derives from a potential, as given by (2.39), the homogeneous set of Maxwell's equations (2.35) is automatically satisfied. Notice that the definition of $F_{\mu\nu}$ in terms of A_μ is left unchanged if one rescales $A_\mu \to A_\mu + \partial_\mu \lambda$ with λ an arbitrary function. Such transformations are known as gauge transformations and leave Maxwell's theory unchanged.

Further reading

The treatment of the material in this chapter is barely what is needed to work with special relativity, tensor notation, and electromagnetism. The book and online notes by Carroll (2003) follow a similar presentation and can be used for a more in-depth study. The book by Rindler (1977) also has a good introductory presentation.

Problems

1. The pole in the barn paradox. A pole vaulter runs very fast into a barn that is shorter than the runner's pole. Since Lorentz transformations imply lengths of moving objects contract, an attendant standing near the door claims the pole is shorter than the barn and closes the door once the runner has entered it. So it appears that a pole longer than the barn has been fitted into it. Is this possible? Sort the situation out using a space-time diagram.
2. Show that $\sum_\mu D^{\mu\mu}$ and $\sum_\mu D_{\mu\mu}$ are not invariant under Lorentz transformations but $\sum_\mu D^\mu{}_\mu$ is.
3. Show explicitly that the boost (2.11) indeed leaves the distance (2.8) invariant.
4. Show that $\epsilon^{\sigma\alpha\beta\gamma} \partial_\gamma F_{\alpha\beta} = 0$ is equivalent to half of the Maxwell equations.
5. Obtain the magnetic field due to an current by considering the electric field of an infinite charged wire, and Lorentz boosting it into a frame

in which the wire is moving along its length. Does it yield the usual result?

6. Show that the homogeneous Maxwell equations (2.35) are automatically satisfied if the field tensor is related to a vector potential as in (2.39).

7. Write the Lorentz force law, $\vec{F} = q\vec{E} + q\vec{v} \times \vec{B}$ in covariant form.

3
Some elements of general relativity

3.1 Introduction

One of the main contributions of Newtonian physics was to establish a simple law to describe gravity. The latter is the most universal force in nature, all physical objects experience it. Newton's law, that establishes that the gravitational force between two point masses is proportional to the masses and inversely proportional to the distance squared, successfully explained the solar system, the motion of projectiles and of everyday objects. However, it is clear that Newton's gravity cannot be the final picture since it is clearly incompatible with special relativity. The force between two masses acts instantaneously, there is no delay in propagation. For example, if one has two masses and increases one of them, the other will experience an increased force immediately, no matter how far away it is.

Gravity also has another privileged characteristic unlike that of any other force. The mass that appears in Newton's second law representing the inertia of a particle to be accelerated is also the same mass that appears in Newton's law of gravitational attraction. As a consequence of that, given a gravitational field, all masses experience the same acceleration in it. This is sometimes called *the equivalence principle*. This would not be the case in an electromagnetic field, where given a field, different charges suffer different accelerations. This fact allows us to mimic (at least locally) gravitational effects by going to an accelerated frame. This is the origin of the "Einstein elevator" thought experiment. If someone is in a small elevator that is free falling in the Earth's gravitational field or an elevator that is floating in outer space, no local experiment carried out inside the elevator could decide which is which. Similarly an elevator fixed on the ground on Earth and an elevator in outer space that is accelerated with an acceleration of 9.8 m/s cannot be distinguished by local experiments. The need for locality is that, if the elevator is large, one could drop two objects and in an accelerating frame they would drop in parallel (neglecting the gravitational attraction between the two masses) whereas on the Earth they would gradually move together since the gravitational field points out radially. What we conclude with all this is that if one includes accelerating frames of reference, one is de facto including gravity. The equivalence of inertial and gravitational mass has been tested with enormous accuracy (10^{-13}) by using torsion balances and studying the influence of the gravity of the Earth, Sun, and galactic centers using different masses in the balance (Schlamminger *et al.* 2008).

The fact that accelerated frames are related with curved geometries can also be motivated by a thought experiment also considered by Ehrenfest. Suppose you are in a carousel that is not moving and you measure its diameter and its circumference using a ruler. You will get that they are proportional and the proportionality factor is π. Now suppose the carousel starts to accelerate and it achieves speeds comparable to the speed of light. If you now try to measure the circumference, the ruler will Lorentz contract and you will measure a longer circumference than before. When you measure the diameter, as the ruler is transverse to the motion, it will not Lorentz contract. You will conclude that the value of π has changed. Kaluza's explanation of this apparent paradox is that the relation between diameter and circumference is π in flat spaces. By going to an accelerating frame you have made space look curved, at least in this sense. One can build similar thought experiments with clocks that suggest that space-time is curved. Unfortunately, as we will see, these arguments are naive. One cannot make a flat space curved just by changing reference frames. One of the major mathematical challenges in dealing with curved space-times is how to recognize true curvature from coordinate dependent notions. We will deal with this in Section 3.3. The Ehrenfest carousel also involves several other subtleties related to rotating frames that make its interpretation difficult, a good summary is present in the relevant Wikipedia article.

Summarizing, general relativity will be a theory where space-time will be generically curved and in which the physics will be invariant under any changes of reference frame. It will also be a theory of gravity, where the latter will not be represented by a force, but by the curvature of space-time. As such it is unique among the fundamental interactions, since all others are described by a force living in a background space-time. As we discussed in the introduction, we will not attempt to cover in detail general relativity, just introduce some basic elements.

3.2   General coordinates and vectors

As we just stated, general relativity is a theory of gravity, but instead of describing gravity as a force, it describes it as a deformation of space-time. To deal with this theory we therefore need to learn a bit of how to handle curved geometries. The first observation is that the metric in a curved space-time is not given anymore by the Minkowski metric $\eta_{\mu\nu}$. Generically, the distance between two infinitesimally close points in a curved space-time is given by,

$$ds^2 = g_{\mu\nu} dx^\mu dx^\nu, \tag{3.1}$$

with $g_{\mu\nu}$ a symmetric matrix possibly dependent on the coordinates. For instance, in the vicinity of the Earth, the metric looks approximately like,

$$ds^2 = -\left(1 + 2\Phi(\vec{x})\right) dt^2 + \left(1 - 2\Phi(\vec{x})\right)\left(dx^2 + dy^2 + dz^2\right) \tag{3.2}$$

where $\Phi(\vec{x})$ is the Newtonian potential at the point one is evaluating the metric. In units in which $c = 1$ the gravitational potential at the surface of the Earth is about 10^{-10}, which explains why we are usually justified in ignoring space-time curvature. However, that tiny deviation from flatness is responsible for all objects falling towards the Earth at $9.8\,m/s^2$!

An immediate complication that arises when one deals with curved geometries is that there is no concept of Cartesian coordinates (or generically, there exist no preferred set of coordinates). In fact there is no concept of straight line. One is forced to use what one normally refers to as "curvilinear coordinates." But this adds several complications. First of all, it makes it harder to tell if a space-time is flat or not. For instance, consider the metric,

$$ds^2 = -dt^2 + dr^2 + r^2\left(d\theta^2 + \sin^2\theta d\varphi^2\right). \tag{3.3}$$

We would all recognize that (apart from the dt^2 bit) this is just the usual expression for distance in spherical coordinates. This space-time is indeed flat. But the metric is not $\eta_{\mu\nu}$, in fact it has a spatial dependence. Worse, there even are points $(\theta = 0, \pi)$ where one of the metric coefficients seems to vanish! So this is the first challenge that curved geometries pose to us: how do we know if a geometry is curved? One could argue that in a flat geometry one can always choose coordinates such that the metric is $\eta_{\mu\nu}$, and therefore, if you are handed a metric you prove it is flat if you can change coordinates to those in which it is simply $\eta_{\mu\nu}$. But that generally turns out to be an incredibly challenging task! To explicitly construct the coordinate transformation requires the solving of a set of coupled partial differential equations, which generically one does not know how to solve. That is, if one is given a set of coordinates x^μ in which the metric is $g_{\mu\nu}$ to find coordinates $x^{\mu'}$ in which the metric is $\eta_{\mu'\nu'}$ one needs to solve,

$$g_{\mu\nu}dx^\mu dx^\nu = \eta_{\mu'\nu'}dx^{\mu'}dx^{\mu'} \tag{3.4}$$

and by this it is meant to solve $x^{\mu'}$ viewed as a function of x^μ. The resulting equations are non-linear and generically with non-constant coefficients, that is why this is such a hard problem (historically it was given the name of "the equivalence problem" and haunted Gauss and many other famous mathematicians). The bottom line is: one needs to develop a better framework to tell if a metric is curved.

The use of curvilinear coordinates which are forced upon us in curved space-time creates an additional complication: how to do vector analysis. First of all we need to reconsider our definition of vectors. A curved space-time behaves locally at every point like a vector space. The picture that emerges is shown in Fig. 3.1. So at each point in the space one indeed has a vector space. One can take its elements, add them, compute their scalar products, and multiply them times a scalar, just like in any vector space. But the first warning is that one is not supposed to mix vectors at different points in the space since they belong in different vector spaces. In ordinary vector calculus we are accustomed to "sliding vectors around" to add them, etc. This is only possible in flat spaces. In curved ones we

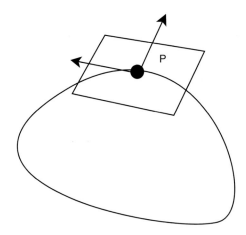

Fig. 3.1 *At a given point P, curved space-times behave like vector spaces.*

will have to be more careful. An immediate problem crops up in the definition of a derivative of a vector. The notion of derivative involves comparing quantities in nearby points. But we just said that we cannot mix vectors at different points. This has a practical implication. In flat space-time in Cartesian coordinates we had that a vector transformed as the coordinates $A^{\mu'} = \Lambda^{\mu'}{}_{\mu} A^{\mu}$ with the matrices Λ being only dependent on velocities. In curvilinear coordinates the relationship between the old and new coordinates is in general a complicated non-linear function $x^{\mu'} = x^{\mu'}(x^{\mu})$. However, the transformation law for the differential of the coordinates is linear,

$$dx^{\mu'} = \Lambda^{\mu'}{}_{\mu} dx^{\mu} \quad \text{with} \quad \Lambda^{\mu'}{}_{\mu} = \frac{\partial x^{\mu'}}{\partial x^{\mu}}. \tag{3.5}$$

Since, as we argued, one can only think of vectors at a point, the differential of the coordinates is a more natural thing to use as a guideline (since its value is local) than the coordinates themselves. We will therefore define a vector in curvilinear coordinates as a set of numbers that transform as the coordinate differentials. Vectors with lower indices transform with the inverse matrix, that is,

$$A_{\mu'} = \Lambda_{\mu'}{}^{\mu} A_{\mu} \quad \text{with} \quad \Lambda_{\mu'}{}^{\mu} = \frac{\partial x^{\mu}}{\partial x^{\mu'}}. \tag{3.6}$$

Let us return to the issue of how to take derivatives of a vector. Just by taking derivatives with respect to the coordinates of the above expression, and for the moment assuming the Λ matrices are coordinate independent, we have,

$$\partial_{\nu} A_{\mu'} = \partial_{\nu}(\Lambda_{\mu'}{}^{\mu} A_{\mu}) = \Lambda_{\mu'}{}^{\mu} \partial_{\nu} A_{\mu}. \tag{3.7}$$

The left and right members look funny because we are mixing a quantity in the new coordinate system x' with derivatives in the old coordinate system x. This can be corrected by noting that since, in Cartesian coordinates, $x^{\mu'} = \Lambda^{\mu'}{}_{\mu} x^{\mu}$ then $dx^{\mu'} = \Lambda^{\mu'}{}_{\mu} dx^{\mu}$ or,

$$\frac{\partial x^{\mu}}{\partial x^{\mu'}} = \Lambda^{\mu}_{\mu'}, \tag{3.8}$$

and therefore applying the chain rule to (3.7) we get,

$$\partial_{\nu'} A_{\mu'} = \Lambda_{\nu'}{}^{\nu} \Lambda_{\mu'}{}^{\mu} \partial_{\nu} A_{\mu}. \tag{3.9}$$

That is, the derivative of a vector transforms as a tensor. Crucial in this derivation is that the derivative sailed through the matrix Λ in the last identity in (3.7) since Λ does not depend on the coordinates. This is true in Cartesian coordinates. When one is dealing with curvilinear coordinates, this is not true anymore. The matrices implementing coordinate transformations generically depend on the coordinates. The partial derivative of a vector fails to be a tensor, one picks inhomogeneous terms, depending on the transformation law, that do not vanish and are coordinate dependent. Being able to take derivatives is crucial for setting up physical theories, so this needs to be corrected.

The way to correct it is by defining a new type of derivative called a *covariant derivative*. The covariant derivative differs from the ordinary partial derivative in a term that is linear in the vector we are discussing (otherwise the derivative would fail to be a linear operator). It is denoted by ∇_μ,

$$\nabla_\mu A^\nu = \partial_\mu A^\nu + \Gamma^\nu_{\mu\lambda} A^\lambda. \tag{3.10}$$

What are the three indexed objects $\Gamma^\lambda_{\mu\nu}$? They are called a *connection*, because they allow us to "connect" neighboring points on the space-time in order to compute derivatives.

So where does one get the connection? The connection in general is an additional element that needs to be provided by whomever is providing you with the curved space-time, just like they provided you with the metric. One cannot do vector analysis if only supplied with a metric, one needs this additional element: the connection. It turns out that if one demands that the connection be symmetric in its lower indices $\Gamma^\lambda_{\mu\nu} = \Gamma^\lambda_{\nu\mu}$ and if one demands that the connection be such that the covariant derivative of the metric is zero, $\nabla_\sigma g_{\mu\nu} = 0$, then the connection is completely and uniquely determined by the metric[1]. Note that these are choices, generally it is not determined by the metric. But it turns out that the geometries that are useful in general relativity do satisfy those requirements, they are called Riemannian geometries. It just happens that nature works that way with gravity, but it could have been different. People tried to exploit more general geometries to unify gravity with

---

[1]In the technical jargon the symmetry of the connection is called "torsion free" and the annihilation of the metric by the covariant derivative is called "the connection is metric-compatible."

electromagnetism many years ago, but such efforts proved unsuccessful in the end (Goenner 2004).

The expression for the connection in terms of the metric in a Riemannian geometry is,

$$\Gamma^{\lambda}_{\mu\nu} = \frac{1}{2}g^{\lambda\rho}\left(\partial_{\mu}g_{\rho\nu} + \partial_{\nu}g_{\rho\mu} - \partial_{\rho}g_{\mu\nu}\right). \tag{3.11}$$

If you are curious about where this comes from, one takes $\nabla_{\sigma}g_{\mu\nu} = 0$, writes it explicitly, and cycles its indices and sums. This formula is known as Christoffel formula (or archaically "Christoffel symbol"). Notice that this formula gives automatically vanishing connection coefficients if the metric is Minkowski and one is in Cartesian coordinates. But it will give non-trivial coefficients if the metric is flat space, say, in spherical coordinates. These extra terms are related to those extra terms that you have already encountered in past courses when you discussed gradient, curl, and divergence in curvilinear coordinates, for instance in electromagnetism. Notice that the connection is not a tensor: it can vanish in one coordinate system and does not vanish in another. This makes sense: we could not have fixed the problem that the derivative of a vector is not a tensor by adding a bit to its definition that is a tensor. Inevitably what we added had to fail to be a tensor itself.

3.3 Curvature

We still do not have a satisfactory definition of curvature. True, if the connection coefficients vanish identically everywhere then the metric is flat. But otherwise we do not know. Curvature, in fact, is not a local concept. It turns out one can always choose locally (at one point) coordinates where the metric is Minkowski and the connection coefficients vanish. This is what one operationally did in the Einstein elevator near the Earth when one allowed it to free fall: locally one could not see any effects of gravity. Due to this property we can set up local coordinates where things look Minkowskian. So how do we know we are in a curved space? The answer is necessarily non-local, that is you need to "take a walk" around your space (or in infinitesimal calculations, one needs to involve at least second derivatives). Consider that you and a friend try to establish if the Earth is flat or not. To do so, you grab a straight piece of metal and walk around the Earth while trying to keep the straight piece of metal as parallel to itself as you can as you move (and tangent to the

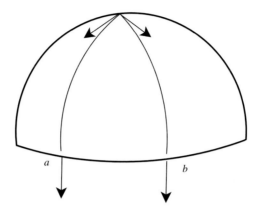

Fig. 3.2 *One can determine if a surface is curved by attempting to transport a straight line around it in a closed circuit and seeing if it comes back pointing in the same direction. Here starting in the North Pole, go to point a on the equator, then to point b, and then return to the North Pole to find the straight line points at a different angle. The angle is bigger the bigger the area enclosed by the circuit.*

Earth, we are trying to ascertain the curvature of the surface of the Earth as a two-dimensional space and you and your friend as having negligible height, since otherwise we would be making measurements outside the space of interest). Suppose, as in Fig. 3.2, that you and your friend start at the North Pole and walk down towards the equator. Your metal rod will point southward at the equator at point a. Suppose you now go to point b, again keeping the rod as parallel to itself as possible and you then backtrack to the North Pole. When you arrive at the North Pole you'll realize that the rod is at an angle with respect to the orientation you picked at departure. The angle will be bigger the bigger the area enclosed by your trajectory (if you go down and up again on the same meridian, for instance, the angle is unchanged as you enclose zero area). This notion of curvature is a bit different from the colloquial usage of the word. Consider a circle. If one had one-dimensional living organisms on the circle, and they tried the above mentioned method to determine if they are living in a flat or curved space and attempted to carry a metal rod around the circle, they would conclude that it is flat. That is, the metal rod will depart and arrive at a point pointing in the same direction after going around the circle. The same is true for a cylinder. So are we missing something with this notion of curvature? The answer is yes. The notion of curvature we have introduced is the best that can be done from within a space. It is called *intrinsic* curvature. There is no mechanism that one can devise for the poor creatures living on a circle that will tell them that it is not flat using measurements they can make within the circle. There exists another notion of curvature which pertains to how a space relates to a higher dimensional space and observers external to it (like us observing the creatures living on a circle). Such a notion is called *extrinsic* curvature and we will return to it later on in this chapter.

Going back to the intrinsic curvature, we need to translate the above described non-local notion of the curvature to something useful for differential calculus purposes. One possibility is to consider moving around an infinitesimally small loop. The change in the direction of the vector would be small too. If we build an infinitesimal closed loop like the one shown in Fig. 3.3 we could consider moving from a to b via point c or via point d. In one case we would advance first through dx^μ, and then through dx^ν, and in the other the order would be reversed. This leads us to consider the double derivative (once in the x^μ direction, once in the x^ν direction) to displace from a to b. In the other case the order of the derivatives should be reversed. The expression we want to consider, therefore, is $(\nabla_\mu \nabla_\nu - \nabla_\nu \nabla_\mu) V^\lambda$. Notice that one is forced to involve an object with at least one index like a vector V^λ, since covariant derivatives on a scalar are just ordinary partial derivatives and therefore commute (at least when the connection is symmetric as we will assume). Now, obviously the rate of change in the direction of the vector should be proportional to the vector itself. And as we see from the previous expression, it should produce an object with three indices μ, ν, and λ. That means that the

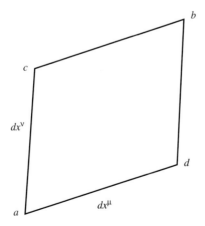

Fig. 3.3 *The infinitesimal version of taking a circuit around a curved space.*

proportionality factor ought to be an object with four indices. Such an object is usually called the Riemann tensor or curvature tensor,

$$\left(\nabla_\mu \nabla_\nu - \nabla_\nu \nabla_\mu\right) V^\lambda = R^\lambda{}_{\rho\mu\nu} V^\rho. \tag{3.12}$$

An object with four indices may appear intimidating, but it is rather natural in this context: we need to produce as a result a vector (one index) but the result will depend on how the infinitesimal element of area is oriented (two indices) and on the vector we start from (another index). If the curvature tensor is zero the derivatives commute. Since we can construct a finite closed circuit by combining infinitesimal ones (at least if no non-trivial topological features are involved), if the curvature tensor vanishes at all points, then we are assured the space is flat. We have therefore succeeded in characterizing in an unambiguous way the idea of curvature. It does not matter how crazy the coordinates one is working with, this definition will work because a tensor that vanishes in one coordinate system vanishes in all coordinate systems. Notice that the connection coefficients in general will be non-zero and the covariant derivatives will be complicated, but nevertheless if the space is flat the curvature will be zero. You are capable of working out, if you wish, the explicit expression of the curvature tensor in terms of the connection coefficients. It just involves writing out the expression for the covariant derivatives and playing around a bit. Since we will not use this formula in the book, we omit it.

At this point it is good to introduce an additional tensor operation we will find useful. When one has a tensor with a pair of indices one up and one down, one can sum over them and the result is a tensor with two indices less (you saw a particular case in Problem 2.2). Such an operation is called "contraction" and is the generalization to this context of the idea of the trace one uses in matrices. If one views a matrix as a tensor with an index up and an index down, summing over the indices is nothing more than summing the diagonal elements of the matrix, that is taking the trace. If one wants to contract two indices that are both up or down, one must raise one or lower one using the metric as we discussed before. When a tensor has many indices there are several possible "traces" (contractions) that one can introduce. In the case of the Riemann tensor, common ones are the Ricci tensor,

$$R_{\mu\nu} \equiv R^\lambda{}_{\mu\lambda\nu} \tag{3.13}$$

and the curvature scalar,

$$R = R_{\mu\nu} g^{\mu\nu}. \tag{3.14}$$

3.4 The Einstein equations and some of their solutions

Up to now all this has been a mathematical construction. The Einstein equations are the physical equations that determine which geometry occurs in nature. They state that a linear combination of the components of the Ricci tensor are determined by the amount of energy and stress present in matter. Specifically,

$$R_{\mu\nu} - \frac{1}{2} g_{\mu\nu} R = 8\pi G T_{\mu\nu}. \tag{3.15}$$

In this equation $T_{\mu\nu}$ is constructed from matter fields and is known as a stress-energy tensor or energy-momentum tensor and G is Newton's constant. This equation is similar in spirit to the equation that determines the Newtonian potential $\nabla^2 \phi = 4\pi G \rho$, where ϕ is the gravitational potential and ρ is the density of matter. Just like the left-hand side of that equation has second derivatives of ϕ, the left-hand side of the Einstein equations has second derivatives of the metric, which plays the role of gravitational potential. There are, however, important differences. The Newtonian equation only has second *spatial* derivatives. That means that any change in time on the right-hand side instantaneously propagates to the left-hand side. Newton's theory therefore clearly violates special relativity. The Einstein equations contain both space and time derivatives. This means that changes in the matter properties do not propagate instantaneously to the metric. This respects the spirit of relativity, where one knows that nothing can propagate instantaneously.

At this point it may be useful to consider some examples of solutions of the Einstein equations. Of course, flat space is a solution of the vacuum equations (when $T_{\mu\nu} = 0$). But there exist other solutions in vacuum. An important solution is the *general* solution in vacuum with spherical symmetry, known as the Schwarzschild solution,

$$ds^2 = -\left(1 - \frac{2GM}{c^2 r}\right) dt^2 + \left(1 - \frac{2GM}{c^2 r}\right)^{-1} dr^2 + r^2 \left(d\theta^2 + \sin^2 \theta d\varphi^2\right). \tag{3.16}$$

This solution plays an analogous role to the potential $\Phi = GM/r$ in Newton's theory. Just like that potential is a solution of $\nabla^2\Phi = 0$, the Schwarzschild solution is a solution in vacuum. Unfortunately its simple appearance is quite deceptive. The solution was found by Karl Schwarzschild in 1916 while fighting on the Russian front of World War I and also while gravely ill (he died a few months later). So the discovery of the solution was extremely meritorious (Einstein had considered his equations too hard to solve in closed form, preferring an approximate solution, and was surprised when he heard of Schwarzschild's result). The proper interpretation of the space-time solution it represents was only found many years later by Kruskal (1960) and Szekeres (1960). It is quite remarkable that the best minds in physics of the twentieth century, Einstein, Weyl, Eddington, and others, were baffled by the solution and died without knowing its true meaning. The reason they were all baffled was they managed to confuse what are coordinate dependent effects with true physical effects. We do not need to concern ourselves here with other peoples' confusions (which in the mind of many amateur physicists still persist today!), so we will just state our modern understanding of the metric. A detailed discussion of all the properties of the space-time the solution represents is beyond the scope of this text, but we will list a few. For values of r much larger than M (in units in which $c = G = 1$) space-time is approximately flat and ordinary looking. However, things change dramatically close to $r = 2M$. First of all, at that point the metric has a singularity, which it turns out is not a true singularity, one can actually find coordinates where the metric is regular at $r = 2M$, and that is what Kruskal and Szekeres did. Careful studies indicate that the surface $r = 2M$ is actually not a time-like surface at all but it is null. That is an object at $r = 2M$ is moving at the speed of light even though it remains at $r = 2M$! The consequence of this is that any object that penetrates closer to $r = 0$ than $r = 2M$ cannot get back out unless it moves faster than light. There exists what is usually called a *trapped region* in the space-time. This is what is normally called a *black hole*. The surface at $r = 2M$ is called the *event horizon*. In the region interior to $r = 2M$ the sign of the dt^2 and dr^2 coefficients change. This indicates that the variable r behaves like a "time." Just like outside $r = 2M$ one "inevitably marches into the future" as t increases, inside $r = 2M$ one inevitably marches towards $r = 0$, which is not a point at all but actually a space-like surface that lies to the future of all points interior to the black hole. That is, if you fall into a black hole you inevitably end up at $r = 0$. At that point there is a genuine singularity that cannot be removed by changing coordinates. Close to the singularity the values of all non-vanishing scalars you can build by taking contractions of the curvature tensor diverge. An object approaching that region would be ripped apart by gravitational tidal forces. Black holes are expected to form when stars exhaust the nuclear fuel that, when burned, produces radiation pressure that keeps their surface buoyant against the gravitational self-attraction of the matter constituting the star. When the fuel is exhausted there is no pressure and the surface starts to collapse. Eventually it will cross the $r = 2M$ surface and all the matter in the star will be trapped in the black hole.

Another solution of interest is the Friedmann–Robertson–Walker (FRW) cosmology

$$ds^2 = -dt^2 + a(t)^2 \left(dx^2 + dy^2 + dz^2\right), \tag{3.17}$$

where $a(t)$ is determined by the Einstein equations depending on the matter content. This solution represents a space-time that has spatial sections that are homogeneous (there is no dependence on x, y, z) and isotropic (all three directions are treated equally). We believe that our universe today is very close to homogeneous on large scales. This is called the Copernican principle, which states that our position in the universe is not privileged and therefore it should look the same way at all other points at the same cosmic time. From our vantage point we observe that at large scales the universe is very approximately isotropic. For instance, the cosmic microwave radiation background has a temperature of $2.72°$Kelvin that is isotropic up to one part in 10^5. So this metric could approximate our current universe reasonably well. We allow for spatial distances to change with time. It is not a solution of the vacuum equations (unless a is a constant, in which case one has a flat space-time) but it solves the equations coupled to various forms of matter depending on the functional form of $a(t)$. For instance, if $T_{\mu\nu}$ has as only component $T_{00} = \rho(t)$ it represents a gas of particles of density ρ that are not moving. This is actually not a bad approximation to the matter in our universe today. In a gas of moving particles there would be other components of $T_{\mu\nu}$ that are non-vanishing that are proportional to the velocities, but since we are using units where $c = 1$ the velocities, say of galaxies, are very small numbers. In that case $a(t) = t^{2/3}$. We observe that at $t = 0$ there seems to be a problem. Indeed there is a singularity that cannot be removed by changing coordinates. Again curvatures blow up as one approaches the singularity. This is the Big Bang that most modern cosmology theories predict as the origin of our universe. We will see that loop quantum gravity, when applied to cosmologies, indeed eliminates this singularity, but nevertheless there is a region of large curvature in the space-time close to where classical general relativity predicts the Big Bang occurred. Those high curvatures imply any matter present is at high temperature. The radiation emanating from matter in that era is what constitutes the cosmic microwave background today. The low temperature we observe for it is due to the fact that between the moment of emission of the radiation and today the universe has expanded and the radiation cooled.

A couple of extra remarks about FRW. First of all, what we presented here is what is known as the "flat" model. It is clear that a space being homogeneous does not necessarily mean that its spatial sections are necessarily flat. For instance a sphere is clearly homogeneous and curved. A hyperboloid is another example of a homogeneous space. One can easily generalize the metric (3.17) to homogeneous but curved spatial sections, but we will not need it in this book. A second remark is that in the cosmological context a different kind of "matter" is sometimes considered called a *cosmological constant*. For such matter $T_{\mu\nu} = -\Lambda g_{\mu\nu}/(8\pi G)$ with Λ a constant. The origin of such matter is a historical accident. When Einstein first tried to solve his equations for the homogeneous and isotropic case he readily noted, as we did above, that the universe expands or contracts. However, this was in 1917 and at the time it was widely believed that the universe was static. Einstein noted that if instead of using as the left-hand side of the field equations $R_{\mu\nu} - g_{\mu\nu}R/2$, one used $R_{\mu\nu} - g_{\mu\nu}R/2 + \Lambda g_{\mu\nu}$, the theory was still consistent but admitted static universes as solutions. When it was later realized that the universe expands, Einstein called the introduction of the cosmological constant "the biggest blunder of my life" (Gamow 1970). In modern cosmology, observations of distant supernovae suggest that the cosmological constant is non-zero. On the other hand, the energy momentum tensor for the vacuum state of a quantum field theory has exactly the form of a cosmological constant, which leads some to believe that it could be its origin. Unfortunately, a naive calculation shows that the value predicted by quantum field theory is 120 orders of magnitude too big. It is expected that a quantum theory of gravity may shed light on this problem. Some proposals exist, but none has achieved wide consensus among theorists to date (for an alternative viewpoint see Bianchi and Rovelli (2010)).

3.5   Diffeomorphisms

At this point it is a good idea to introduce the concept of a *manifold*. A manifold is a topological set (a set in which one can define neighborhoods of elements and take limits) that is mappable (at least in local patches) to R^n, that is, one can set up coordinates on it to distinguish its elements. For instance, a collection of chairs is not a manifold since there is no notion of neighborhood. But a tabletop is. A tabletop with a nail sticking out of it is not (the tabletop may be mappable to R^2 and the nail to R but the neighborhood of the point where the nail enters the tabletop cannot be mapped to R^n). The equations of general relativity are invariant under general coordinate transformations. There are two ways to deal with coordinate transformations. One of them, most commonly considered in physics, is to keep the points of the manifold fixed and change the mapping into R^n. This is a "change of coordinates." But there is another possible perspective, and that is to keep the map to R^n fixed and move around the points in the manifold. The way we "move around points" is through a type of mathematical map called *diffeomorphisms*. A diffeomorphism is a one-to-one map that given a point p in the manifold, maps it to a new point $\Phi(p)$, and does so respecting topological notions of "proximity", so if one was able to take derivatives of quantities at point p before the diffeomorphism, one can still do it after it. This latter perspective on coordinate transformations is called the *active view* and the more common one in physics contexts is called the *passive view*. The active view is more natural in the context of a theory of gravity like general relativity. The latter is a theory of a metric that lives on a manifold. On such a manifold there are no preferred points until one introduces the metric. Each point has the same status as any other. Therefore it ought to be possible to "move the points around" without any consequence for the theory. This, in a nutshell, is the reason why any theory of gravity of geometric origin has to be invariant under diffeomorphisms (or if you want, under general coordinate transformations). It just reflects that the theory has no preferred background structures that can tell apart one point from another on the manifold. This is a concept that takes some work to get used to because all other physics one has normally dealt with do not have this type of invariance.

Diffeomorphisms "move around" not only the points of the manifold but everything that is sitting on it. To understand this point better it is good to discuss first the more general concept of a *map* between sets and how maps move around objects. Suppose you have two different sets M and N and you have a map Φ that assigns to each point in M a point in N. Notice that M and N don't even have to have the same dimensionality in principle, we will assume they do just for simplicity. So if p is a point in M then $q = \Phi(p)$ is a point in N. We will call this operation a *push forward* of the point p into N. Now, suppose you have a function f that assigns a real number to each point in N. Such a function in principle does not know how to act in M. But we can use our map to teach it how to act in M. We define a new function $\Phi^*(f)$ in the following way $\Phi^*(f)(p) \equiv f(\Phi(p)) = f(q)$ and it is clear that such a function acts on M. Using the map in this way is called the *pull back* of the function f from N to M.

What about vectors and tensors, can we push forward and pull back those too? A vector can be thought of as an operator acting on functions. How? Given a vector V^μ, one can define the derivative of a function f along such vector $V^\mu \partial_\mu f$ and the result is a function. So say one has a vector acting on M. One can define an operator in N acting on functions by considering the *push forward* of the vector defined like this $V^\mu \partial_\mu(\Phi^*(f))$. That is, the vector acts on the pull back of the function.

This sounds a bit abstract. Can we find out the components of such a vector? We can. Consider,

$$(\phi V)^\mu \partial_\mu f = V^\mu \partial_\mu(\phi^*(f)) = V^\mu \partial_\mu(f \circ \phi). \tag{3.18}$$

To compute this explicitly we need to introduce coordinates. Say the coordinates in M are called x^μ and the coordinates in N are called y^μ. Then the last member of the previous identity can be written as,

$$V^\mu \frac{\partial(f \circ \phi)}{\partial x^\mu} = V^\mu \frac{\partial y^\nu}{\partial x^\mu} \frac{\partial f}{\partial y^\nu}. \tag{3.19}$$

To write the above expression we need to assume that the map is smooth, since we are computing derivatives, and we will assume it is invertible so one can go back and forth between x and y. Such a map is called a diffeomorphism. Suddenly, the abstract has become very concrete: the effect of the diffeomorphism on the vector is to multiply its components times the matrix $\partial y^\nu / \partial x^\mu$. But that is the matrix of coordinate transformations! So if we had assumed that the two manifolds were the same, the practical effect of the diffeomorphism is the same as a coordinate transformation in the manifold, as we were anticipating from the beginning.

It should be clear from the above discussion that certain objects can be pushed forward and certain others pulled back but the converse is generically not true. Unless the map is invertible. Then one can repeat the above construction swapping N and M just by considering the inverse of the map, and define push forwards and pull backs for all quantities. Since diffeomorphisms are invertible maps we will be able to pull back and push forward via diffeomorphisms any object we like.

One last thing we need to consider is the concept of a one-parameter family of diffeomorphisms Φ_t. This is a family of maps such that for $t = 0$ one has the identity map and the maps move the points "farther away" for bigger values of the parameter t. Starting from a given point p, the continuous family of diffeomorphisms $\Phi_t(p)$ generates a curve on the manifold. We can compute the tangent vector to such a curve $V^\mu = dx^\mu / dt$.

The notion of diffeomorphism can actually be used to define a new derivative of vectors and tensors. It will have the property of being a vector or tensor without the need to introduce extra structures. It is called the Lie derivative. Given a one parameter family of diffeomorphisms the Lie derivative of a tensor T along a vector is given by

$$\mathcal{L}_V T(p) = \lim_{t \to 0} \frac{\phi_t^*(T(\phi_t(p))) - T(p)}{t}. \tag{3.20}$$

So we push forward the point p via a diffeomorphism along a curve whose tangent vector is V^μ, evaluate the tensor at the new point and then pull back the whole thing to p. That way, we can subtract from it $T(p)$, since we are allowed to add and subtract vectors and tensors provided we do so at a given point. It is clear that the result of this operation is a tensor of the same rank as the one we started with.

We will not explicitly derive here the form of the Lie derivative on an arbitrary tensor. In principle by generalizing a bit the results for the vectors we showed above you could work it out, we just quote the formula and refer to the book by Carroll (2003) for further details,

$$\mathcal{L}_V T^{\mu_1\mu_2\cdots\mu_k}{}_{\nu_1\nu_2\cdots\nu_m} = V^\sigma \partial_\sigma T^{\mu_1\mu_2\cdots\mu_k}{}_{\nu_1\nu_2\cdots\nu_m} - \left(\partial_\lambda V^{\mu_1}\right) T^{\lambda\mu_2\cdots\mu_k}{}_{\nu_1\nu_2\cdots\nu_m} \qquad (3.21)$$

$$- \left(\partial_\lambda V^{\mu_2}\right) T^{\mu_1\lambda\cdots\mu_k}{}_{\nu_1\nu_2\cdots\nu_m} \cdots + \left(\partial_{\nu_1} V^\lambda\right) T^{\mu_1\mu_2\cdots\mu_k}{}_{\lambda\nu_2\cdots\nu_m}$$

$$+ \left(\partial_{\nu_2} V^\lambda\right) T^{\mu_1\mu_2\cdots\mu_k}{}_{\nu_1\lambda\cdots\nu_m} + \cdots$$

Lie derivatives will prove useful when discussing symmetries since a symmetry usually means that certain things remain invariant when one performs a diffeomorphism. For instance, a space-time invariant under translations will be given by a metric that has zero Lie derivative along a particular vector field which is tangent to a straight line (it is not heretical to mention a straight line here since we are talking about a particular family of space-times for which straight lines exist!). The same holds for invariance under rotations around an axis, the metric in that case will have zero Lie derivative with respect to a vector that is tangent to a circle.

You may wonder why, if the Lie derivative exists without having to introduce a connection, did we bother with defining the covariant derivative? The latter proved crucial to define the notion of curvature in a coordinate independent way. One cannot use the Lie derivative to do that.

3.6 The $3 + 1$ decomposition

We will eventually be discussing a Hamiltonian formulation of general relativity. If you recall the Hamiltonian formulation of ordinary mechanics, it is given in terms of a set of canonical variables q, p at a given instant of time t. If one is dealing with fields rather than a mechanical system then the canonical variables are functions of position $\phi(x)$ and their canonical momenta are as well $\pi_\phi(x)$, both given at an instant of time. General relativity, as we have discussed it up to now, treats space and time on the same footing. This is not what is done in Hamiltonian formulations. So we will need to break that equal footing in order to discuss general relativity in a Hamiltonian fashion. Some people consider this a significant drawback of the Hamiltonian formulation since it imposes an artificial distinction on the variables of the theory. It is actually slightly worse than that. Although we have not emphasized it, the space-times of general relativity cannot only have arbitrary metrics but also arbitrary topologies. In the Hamiltonian formulation, since one will single out a time-like direction in which things evolve, the topology will come out naturally to be $\Sigma \times R$ where Σ is a three-dimensional manifold (of arbitrary topology) and R is the real line, and it cannot change upon evolution. Some people consider this too restrictive, though it is crucial in order to have a well posed initial value problem. We will see that the Hamiltonian formulation is "smart" in the sense that the framework ends up noticing the arbitrary distinction we introduced and actually compensates for it, making the final result insensitive to our choice of a time-like direction. However, the restrictions on topology remain. But this will have to wait till further chapters.

Here we will therefore assume that space-time has a $\Sigma \times R$ topology and that there is a time-like direction characterized by a vector t^μ whose orbits[2] are a curve parametrized by a parameter t and such that $t =$ constant surfaces are spatial slices Σ. We introduce a vector field n^μ normal to Σ. One can then define a positive definite spatial metric on Σ,

$$q_{ab} \equiv g_{ab} + n_a n_b. \tag{3.22}$$

In the above expression the indices a, b go from 1 to 3. One does not need to identify coordinates in such a way, but it makes things easier if we just view the coordinates $1, 2, 3$ as coordinatizing Σ. Otherwise this is a great place to use the framework we set up when we discussed diffeomorphisms, setting a map between the three-dimensional surface and the four-dimensional space-time and pulling back things appropriately. We can decompose the vector t^μ in components normal and tangential to Σ as, $t^a = N n^a + N^a$. This expression is tricky since the index on t^a really could go from 0 to 3, and so does the index on n^a, but not the one in N^a. So if you want to evaluate the zeroth component of that expression only the first term contributes. The quantities N and N^a are usually known as the *lapse* and the *shift vector*. They can be viewed as a scalar and a vector living in Σ. In terms of these the four-dimensional metric can be written as,

$$ds^2 = \left(-N^2 + N_a N^a\right) dt^2 + 2N_a dt dx^a + q_{ab} dx^a dx^b. \tag{3.23}$$

One can define the *extrinsic curvature*, a notion we mentioned before, as,

$$K_{ab} = \frac{1}{2} \mathcal{L}_n q_{ab}, \tag{3.24}$$

[2]The orbit of a vector is a curve such that its tangent is the vector of interest.

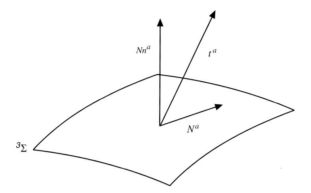

Fig. 3.4 *The vectors involved in the 3+1 decomposition of space-time.*

and it is closely related to the "time derivative" of the metric,

$$\dot{q}_{ab} \equiv \mathcal{L}_t q_{ab} = 2N K_{ab} + \mathcal{L}_{\vec{N}} q_{ab}. \tag{3.25}$$

These are the variables that we will use to construct a Hamiltonian formulation for gravity. The configuration variable will be the three-dimensional metric q_{ab} and its conjugate momentum will be related to the extrinsic curvature K_{ab}. Just like the usual momentum is related to the time derivative of the configuration variable, so is the momentum of the metric. These variables will evolve in time, helping us build, with the lapse and the shift, the space-time metric. The lapse and shift will end up being arbitrary functions, reflecting the freedom to choose four components of the metric to take whatever value we want by choosing a coordinate system.

3.7 Triads

When one works in loop quantum gravity one uses a description of gravity that is slightly different from the one we introduced above. Since one will work in the Hamiltonian picture when one is interested in the spatial part of the metric as we discussed above, let us consider three-dimensional space rather than space-time and introduce a set of three vector fields E_i^a, $i = 1, 2, 3$ that are orthogonal. We can write,

$$q^{ab} = E_i^a E_j^b \delta^{ij}, \tag{3.26}$$

so on the right-hand side we have flat space in coordinates i, j and on the left-hand side we have curved space with coordinates a, b. We therefore see that the E_i^a are all we need to construct the metric. Such quantities are called "triads." So we now have two different types of indices, "space" indices a, b, c that behave like regular vector indices in a curved space, and "internal" indices i, j, k. Both sets of indices range from 1 to 3. However, if one wishes to raise and lower internal indices one should do that with the flat metric δ_{ij}. Such indices can be "contracted" among themselves just like ordinary indices.

If one is going to operate on objects that have internal indices one needs to introduce an appropriate derivative. Say you have an object with just an internal index G^i. We define a derivative on it based on our experience defining the covariant derivative as,

$$D_a G^i = \partial_a G^i + \Gamma_{a}{}^{i}{}_{j} G^j \tag{3.27}$$

and just like in the covariant derivative the components of the *spin connection* $\Gamma_{a}{}^{i}{}_{j}$ have to be specified externally. In order for the derivative of a scalar (for example $G^i G_i$) to be the ordinary partial derivative one must have that

$$D_a G_i = \partial_a G_i - \Gamma_{a}{}^{j}{}_{i} G_j \tag{3.28}$$

and if one needs to take the derivative of an object with mixed indices one uses both connections, that is,

$$D_a E_i^b = \partial_a E_i^b - \Gamma_{a}{}^{j}{}_{i} E_j^b + \Gamma_{ac}^b E_i^c. \tag{3.29}$$

If one demands the convenient property that the derivative annihilate the triads $D_a E_i^b = 0$, then the spin connection gets completely determined by the connection we introduced with the covariant derivatives. It also annihilates the spatial metric. The explicit formula is not very important at this point.

One can define a curvature for the derivative we introduced, pretty much as in the case of the covariant derivative,

$$\left(D_a D_b - D_b D_a\right) G^i = \Omega_{ab}{}^{i}{}_{j} G^j \tag{3.30}$$

where we call $\Omega_{ab}{}^{i}{}_{j}$ the curvature of the connection, and it can be related to the Riemann tensor but, again, the explicit formula will not be used so for reasons of brevity we omit it.

There is one last point of curved space-times that we need to touch upon. This has to do with how to perform integrations on curved manifolds. Suppose one wishes to compute a volume integral in flat space in Cartesian coordinates. The integral is given by

$$V = \int \int \int dx dy dz. \tag{3.31}$$

So the volume is just given by the integral of the coordinate differentials in the region of interest. If one goes to curvilinear coordinates, one needs to include the Jacobian of the derivatives of the new coordinates with respect to the Cartesian coordinates. For instance in spherical coordinates,

$$V = \int \int \int r^2 \sin(\theta) dr d\theta d\varphi. \tag{3.32}$$

So it is not just given by the integral of the coordinate differentials. In general one has that,

$$V = \int \int \int \sqrt{\det(q)} d^3 x \tag{3.33}$$

where one has as an integrand the square root of the determinant of the metric times the coordinate differentials. If one were interested in a volume of space-time then one would integrate in four dimensions and use the square root of minus the determinant of the space-time metric, in order for the square root to be real. The appearance of the determinant of the metric is because multiplied times the coordinate differentials it transforms in exactly the right way under coordinate transformations to keep the integral invariant.

More generally, if one wishes to integrate a scalar along a volume, one has to multiply it times the square root of the determinant of the metric. It is worthwhile introducing a new type of object called *scalar densities* which transform as a scalar times the determinant of the metric. They are sometimes denoted by including a tilde over the relevant quantity. So if one has a scalar function f, one would have,

$$\tilde{f} = \sqrt{\det(q)} f. \tag{3.34}$$

If one considers the product of n such objects one obtains a *density of weight $+n$*, that is, an object that transforms like a scalar times the square root of the determinant of the metric to the n-th power. Sometimes people use multiple tildes on top of quantities to keep track of the density weight, although typographically this obviously has limitations. Notice that density weights can be positive or negative. Negative weights are sometimes denoted by tildes under a quantity. For instance, in Chapter 7 we will encounter the lapse function with density weight -1, denoted as $\underset{\sim}{N}$.

An important example of scalar density is given by the Dirac delta function, since obviously it can be integrated right away in volume. For conventional reasons, however, its density character is not denoted with a tilde. One can also have objects with indices (vectors and tensors) that are densities. They just transform like the respective object (vector or tensor) times the determinant of the metric. An important example of this is the Levi-Civita symbol, which transforms as a fourth rank tensor of density weight one. Of course one cannot integrate a tensor density (the result would not be a tensor). But if one contracts the tensor density with, say, four vectors, one is left with a scalar density and the object can be integrated.

In loop quantum gravity, for reasons we will discuss later, one ends up working with triads of density weight $+1$,

$$\tilde{E}_i^a = \sqrt{\det q}\, E_i^a, \tag{3.35}$$

So, in particular if one simply writes,

$$\tilde{\tilde{q}}^{ab} = \det(q) q^{ab} = \tilde{E}_i^a \tilde{E}_j^b \delta^{ij}, \tag{3.36}$$

one recovers the metric times a factor given by its determinant. It is clear that $\tilde{E}_i^a$ and E_i^a contain the same information, just rearranged.

Further reading

As in the previous chapter, the material presented here is the bare minimum needed to follow the remaining chapters. The elementary aspects of general relativity can be complemented by the online notes and book by Carroll (2003). For the $3 + 1$ decomposition a readable description is in the article by Brown (2011), the book by Kiefer (2006), or the paper by Romano (1993) which also is a good reference for the subject of triads. The original ADM article also makes good reading, it has been recently reprinted (Arnowitt, Deser, and Misner (2008)). As mentioned in the introduction, there are excellent textbooks for undergraduates on general relativity by Hartle (2003) and Schutz (2009), but their treatment differs from the one in this chapter.

Problems

1. Show that the covariant derivative is a linear operator and that it satisfies the Leibniz rule.

2. Given the formula for the covariant derivative of a vector (3.10) V^μ, assume that the covariant derivative of a vector with its index down is given by,

$$\nabla_\mu V_\nu = \partial_\mu V_\nu + \bar{\Gamma}^\lambda_{\mu\nu} V_\lambda. \tag{3.37}$$

Show that for the derivative of a scalar $V_\mu V^\mu$ to be given just by the partial derivative, one has to have that $\bar{\Gamma}^\lambda_{\mu\nu} = -\Gamma^\lambda_{\mu\nu}$. Therefore the covariant derivative of a vector with its index down is given by

$$\nabla_\mu V_\nu = \partial_\mu V_\nu - \Gamma^\lambda_{\mu\nu} V_\lambda. \tag{3.38}$$

3. The general formula for the covariant derivative of a tensor with k upper indices and l lower indices is given by,

$$\nabla_\lambda T^{\mu_1 \ldots \mu_k}_{\nu_1 \ldots \nu_l} = \partial_\lambda T^{\mu_1 \ldots \mu_k}_{\nu_1 \ldots \nu_l} + \Gamma^{\mu_1}_{\lambda\sigma} T^{\sigma \ldots \mu_k}_{\nu_1 \ldots \nu_l} + \ldots + \Gamma^{\mu_1}_{\lambda\sigma} T^{\mu_1 \ldots \sigma}_{\nu_1 \ldots \nu_l} \tag{3.39}$$
$$- \Gamma^\sigma_{\lambda\nu_1} T^{\mu_1 \ldots \mu_k}_{\sigma \ldots \nu_l} - \ldots - \Gamma^\sigma_{\lambda\nu_l} T^{\mu_1 \ldots \mu_k}_{\nu_1 \ldots \sigma}.$$

Show that the covariant derivative of the metric vanishes if the connection is the Christoffel connection.

4. Show that if the connection is symmetric and the covariant derivative of the metric vanishes, the connection is given by the Christoffel symbol using the method suggested in the text.

5. Show that the Lie derivative is a linear operator and that it satisfies the Leibniz rule.

6. Compute $\rho(t)$ for the dust filled FRW model with $a(t) = t^{2/3}$.

7. Starting from the unit normal to a three dimensional spatial hypersurface $n_\mu n^\mu = -1$ and the condition that it be orthogonal to the shift vector, show that the $\ldots tt$ component of the metric is $-N^2 + N_\mu N^\mu$.

8. What is the form of $a(t)$ if $T_{\mu\nu}$ is zero but one has a non-vanishing cosmological constant?

4

Hamiltonian mechanics including constraints and fields

4.1 Usual mechanics in Hamiltonian form

We will consider systems that are characterized by a Lagrangian, which is a function of a set of variables q_i and their time derivatives $\dot{q}_i$. For simple mechanical systems the Lagrangian is usually the difference between the kinetic and the potential energies. But we need to start to think of this framework as much more general as we will soon be applying it to fields and other systems. The remarkable thing is that once you are given the Lagrangian the rest of the techniques work almost exactly the same in all systems. We will not consider the case where the Lagrangian depends explicitly on time, since it arises infrequently when studying fundamental interactions. One then has $L(q_i(t), \dot{q}_i(t))$. From the Lagrangian one performs a *Legendre transform*. To do that we define the *canonical momenta*,

$$p_i(t) \equiv \frac{\partial L}{\partial \dot{q}_i}. \tag{4.1}$$

This relationship is assumed to be invertible, yielding $\dot{q}_i$ as a function of q_i and p_i. One then defines the Hamiltonian as $H(q_i, p_i) \equiv \sum_{i=1}^{N} p_i \dot{q}_i - L$ and this should be written as a function of q_i and p_i only by eliminating all the $\dot{q}_i$'s using the inversion we mentioned above. The space of all pairs p_i, q_i is called *phase space*. The space of the q_i's is called the *configuration space*.

The dynamics of the system is captured by Hamilton's equations,

$$\dot{q}_i = \frac{\partial H}{\partial p_i}, \tag{4.2}$$

$$\dot{p}_i = -\frac{\partial H}{\partial q_i}. \tag{4.3}$$

So for instance, if we were considering a Harmonic oscillator, the Lagrangian would be $L = m\dot{x}^2/2 - m\omega^2 x^2/2$ with ω the frequency. Noting that $p = m\dot{x}$ the Legendre transform is $H = p\dot{x} - L$ and the Hamiltonian takes the form $H = p^2/(2m) + m\omega^2 x^2/2$. The Hamilton equations would yield $\dot{x} = p/m$ and $\dot{p} = -m\omega^2 x$, and immediately $\ddot{x} + \omega^2 x = 0$.

A useful concept is that of the *Poisson bracket*. It is an operation between two functions of the phase space, $f(p_i, q_i)$, $g(p_i, q_i)$, defined as,

$$\{f, g\} \equiv \sum_{i=1}^{N} \frac{\partial f}{\partial q_i} \frac{\partial g}{\partial p_i} - \frac{\partial f}{\partial p_i} \frac{\partial g}{\partial q_i}. \tag{4.4}$$

In terms of the Poisson bracket one has that,

$$\{q_i, p_j\} = \delta_{ij}, \tag{4.5}$$

and you can check that Hamilton's equations can be rewritten as,

$$\dot{q}_i = \{q_i, H\}, \tag{4.6}$$

$$\dot{p}_i = \{p_i, H\}. \tag{4.7}$$

These equations have as a solution a "flow" (or curve) in phase-space given by $q_i(t), p_i(t)$ as a function of the parameter t given initial conditions $q_i(0), p_i(0)$.

4.2 Constraints

Let us turn now to systems with constraints. It is very common in physics to describe a system using more variables than are really needed. For instance the simple pendulum has one degree of freedom, the angle of the pendulum with the vertical. However, one may choose to describe it using x, y coordinates in the plane of motion. In that case, there will be a constraint $x^2 + y^2 = l^2$ with l the length of the pendulum. There are different motivations for considering the use of redundant variables. In some cases one builds a theory without knowing very well which are the true degrees of freedom in advance. It is only a posteriori that one finds out. An example of this is electromagnetism. The electromagnetic field in vacuum has two degrees of freedom (the two possible polarizations of an electromagnetic wave). Yet we usually describe it in terms of the vector potential A_μ which has four components, or in terms of the field tensor $F_{\mu\nu}$ which has six (or equivalently in terms of an electric and a magnetic field). General relativity also has two degrees of freedom, yet we describe it either in terms of a metric which has ten components or a tetrad that has sixteen. So, as we anticipated, the situation is really very common.

Since one commonly describes theories with more variables than are needed, there exist relations between the variables called *constraints*. These are equations involving the variables of phase space that hold at every instant of time in the evolution of the system. If we write the constraints as $\phi(p_i, q_i) = 0$ then one can also say that the quantity $\phi(p_i, q_i)$ is a conserved quantity of the system (albeit one that vanishes).

Conserved quantities O are associated with the presence of symmetries in physical systems. To see this it is a good idea to notice that just like equations (4.6, 4.7) defined a flow in phase space that was just the time evolution of the system, given any quantity function of phase space like our conserved quantity O one can associate a flow with them in phase space,

$$\frac{dq_i}{d\lambda} = \{q_i, O\}, \tag{4.8}$$

$$\frac{dp_i}{d\lambda} = \{p_i, O\}. \tag{4.9}$$

This flow $q(\lambda), p(\lambda)$ is just a mathematical construction without much physical meaning in general. Its solution is a trajectory in phase space. But if the quantity O is conserved by evolution, it means that it is unchanged when subject to the "evolution flow," $\{O, H\}=0$. And conversely, the last equation means that the Hamiltonian of the system is unaffected by the flow generated by O. That means there is a curve in phase space along which you can move and the Hamiltonian of the system (and therefore all its mechanical properties) are unchanged. The presence of such curves is the sign of the presence of a symmetry. What kind of symmetry will depend on the nature of O. For instance, suppose that you have a system that conserves the z-component of the orbital angular momentum $L_z = xp_y - yp_x$. If that is the case we have that,

$$\{H, L_z\} = 0 = x\frac{\partial H}{\partial y} - y\frac{\partial H}{\partial x} + p_x\frac{\partial H}{\partial p_y} - p_y\frac{\partial H}{\partial p_x}. \tag{4.10}$$

It turns out that this equation can be integrated and the solution is $H(p_x^2 + p_y^2, x^2 + y^2, xp_x + yp_y, xp_y - yp_x)$, that is the Hamiltonian can only depend on combinations of coordinates and momenta that are invariant under rotations around the z axis (the first two are obvious, the latter two are the scalar product and cross product of the position and momenta in the x–y plane, which are also invariant under rotations around the z axis).

So, since conserved quantities imply the existence of symmetries in mechanical systems and constraints are conserved quantities, they also imply the presence of symmetries in physical systems. Constraints are usually handled by the method of *Lagrange multipliers*. Suppose you have a system with $2N$ canonical variables and M constraints $\phi_i = 0$, $i = 1 \ldots M$. Then one takes the Hamiltonian and adds to it the constraints each of them multiplied by a (potentially time dependent) *Lagrange multiplier* λ_i,

$$H = H_{\text{orig}} + \sum_{i=1}^{M} \lambda_i \phi_i, \tag{4.11}$$

and this quantity is called *total Hamiltonian*. One then proceeds to write the Hamilton equations just like before. One will have 2N equations for 2N+M unknowns (the canonical variables plus the Lagrange multipliers). One supplements those equations by the M constraints $\phi_i = 0$ to have an equal number of equations and unknowns.

As we mentioned, one does not necessarily realize from the outset when dealing with a system that it has constraints. One is normally handed a Lagrangian that is a function of some set of configuration variables and their time derivatives (and space derivatives if one is dealing with a field theory). Handling a generic system with constraints is complicated and exceeds the scope of this course. The procedure for doing it was introduced by Dirac and is therefore called the *Dirac procedure*. In this book we will only discuss some particular cases that are of interest and are nevertheless reasonably simple to handle without invoking the full Dirac procedure. A telltale sign that one is dealing with a system that has constraints is if one finds that, in the Lagrangian, one of the variables appears but its time derivative is missing. That implies that the variable in question is not really a dynamical variable no matter what it looks like or what it is called, but is a Lagrange multiplier.

Notice that the presence of the Lagrange multipliers in the total Hamiltonian (4.11) implies that if one works out the equations of motion and solves them given some initial data $q_i(0), p_i(0)$, the solution will be dependent on the arbitrary M Lagrange multipliers. This means that the evolution is not uniquely determined by the initial data. This is reasonable, because one can have different evolutions that result in end configurations that are related by a symmetry of the system, and are therefore physically indistinguishable although they may look mathematically different. The underlying physics is unique. The description we have chosen of it is not. To make things a little more concrete, consider an action of a mechanical system, $S = \int dt L(\dot{q}^i, q^i)\, dt$. One has the usual Lagrange equations,

$$\frac{d}{dt}\frac{\partial L}{\partial \dot{q}^i} - \frac{\partial L}{\partial q^i} = 0, \tag{4.12}$$

and computing the time derivative one has,

$$\sum_j \left[\frac{\partial^2 L}{\partial \dot{q}^i \partial \dot{q}^j} \ddot{q}^j + \frac{\partial^2 L}{\partial q^j \partial \dot{q}^i} \dot{q}^j \right] - \frac{\partial L}{\partial q^i} = 0. \tag{4.13}$$

Determining $\ddot{q}^i$ as a function of the initial conditions $\dot{q}^i(0)$ and $q^i(0)$ becomes a problem of linear algebra. For it to have a unique solution, the matrix $\partial^2 L/(\partial \dot{q}^i \partial \dot{q}^j)$ has to be invertible. If it is not, then one can solve the equations as functions of undetermined parameters (more problematic situations can develop as the equations are inhomogeneous, but we will ignore them here). These are the Lagrange multipliers. Similarly, when one wishes to go to the Hamiltonian formulation, one has to invert the equation that defines the canonical momentum,

$$p_j = \frac{\partial L(\dot{q}^i, q^i)}{\partial \dot{q}^j}. \tag{4.14}$$

By the implicit function theorem, the inversion can be carried out only if the determinant of the matrix of second partial derivatives is not zero, the same condition we encountered before. Otherwise one could invert $\dot{q}^i = \dot{q}^i(q^j, p^j, v^\alpha)$ with v^α arbitrary functions of time, related to the Lagrange multipliers, the number of which depends on the number of zero eigenvalues of the matrix of second partial derivatives. There will be constraints in number equal to the number of the arbitrary functions. We will see this in action in the example in next section.

4.3   Field theories

To try to illustrate the previous point, we will study our first field theory, Maxwell theory. The Lagrangian for Maxwell theory is given by

$L = \int d^3x F_{\mu\nu} F^{\mu\nu}/4$. The configuration variables will be the four components of the vector potential A_μ. However, recall that $F_{\mu\nu} = \partial_\mu A_\nu - \partial_\nu A_\mu$. That means that whenever μ is equal to ν, $F_{\mu\nu}$ vanishes. But that means that if we consider the temporal component of the vector potential A_0, such a component will never be accompanied by a time derivative, ∂_0 in this notation. Only spatial derivatives of A_0 appear. Therefore we had assumed incorrectly that it was a configuration variable. It is a Lagrange multiplier! Notice that for the spatial components of the vector potential A_i we do indeed get time derivatives in the Lagrangian.

So we now know that Maxwell theory as we have formulated it is a theory that will have a constraint. Before going further let us talk a bit about how one canonically formulates field theories. This is a subject most students have not covered before. Although covering it in full will require more time than a course like this allows for, the basics are relatively simple. It is really very similar to the Hamiltonian formulation of mechanics, except that wherever one had partial derivatives in mechanics, one now has functional derivatives in field theories. So, for instance, if the definition of the canonical momentum was $p \equiv \partial L/\partial \dot{q}$, in the case of a field theory like Maxwell, if our configuration variable is the spatial portion of the vector potential A_a, (more on this below) then its canonically conjugate momentum is given by $\pi^a \equiv \delta L/\delta \dot{A}_a$. The *functional derivative* is defined as,

$$\frac{\delta L}{\delta \dot{A}_a(x)} \equiv \frac{\partial \tilde{\mathcal{L}}(x)}{\partial \dot{A}_a(x)} - \partial_b \frac{\partial \tilde{\mathcal{L}}(x)}{\partial(\partial_b \dot{A}_a(x))} \tag{4.15}$$

where $\tilde{\mathcal{L}}$ is the *Lagrangian density* in terms of which the Lagrangian is defined as $L = \int d^3x \tilde{\mathcal{L}}$. Incidentally, in ordinary mechanics one has that given action $S = \int dt L(q, \dot{q})$, the Lagrange equations can be simply written in terms of the functional derivative as $\delta S/\delta q = 0$.

We see that the functional derivative takes into account the fact that the functional L depends not only on a variable like $\dot{A}_a$ but may also depend on its spatial derivatives $\partial_b \dot{A}_a$ (in the case of Maxwell theory it does not). The rationale for the above formula is that in a functional like L, because it appears under an integral, one can always eliminate any spatial partial derivative on a given variable by integrating by parts. Since this is just a rearrangement of the expression, the functional derivative should not change. Since integrating by parts (ignoring boundary terms) introduces a minus sign and a derivative on the rest of the expression, that is what the second term in the definition captures. It is natural to consider the vector potential as a configuration variable, but why only its spatial part? It turns out you do not need to assume that. If you take the full vector potential, including its zeroth component and define its canonically conjugate momentum, you will notice that since the action does not involve $\dot{A}_0$ then its canonically conjugate momenta $\pi^0 \equiv \delta L/\delta \dot{A}_0$ vanishes.

Before proceeding with the canonical formulation, let us verify that the proposed Lagrangian indeed gives rise to Maxwell's equations. To do that, we need to compute the functional derivative of the Lagrangian with respect to A_ρ. It helps to write the Lagrangian as,

$$L = \frac{1}{4} \int d^3x \, F_{\mu\nu} F_{\lambda\kappa} \eta^{\mu\lambda} \eta^{\nu\kappa}. \tag{4.16}$$

The functional derivative will act on the F's, in both cases one gets the same contribution, so we focus on one of them and multiply the end result by two. The field tensor $F_{\mu\nu}$ only depends on A_ρ through its derivatives. So in the functional derivative one would get a contribution of the form

$$-\partial_\sigma \left(\frac{\partial \left(\frac{1}{4} F_{\mu\nu} F^{\mu\nu} \right)}{\partial (\partial_\sigma A_\rho)} \right). \tag{4.17}$$

so the total contribution is

$$\frac{\delta L}{\delta A_\rho}(x) = \frac{1}{2} \frac{\partial F_{\mu\nu}(x)}{\partial (\partial_\sigma A_\rho(x))} \partial_\sigma \left(F_{\lambda\kappa} \right) \eta^{\mu\lambda} \eta^{\nu\kappa}, \tag{4.18}$$

and the first partial derivative only contributes when $\mu = \sigma$ and $\nu = \rho$, or when $\nu = \sigma$ and $\mu = \rho$, yielding two Kronecker deltas. The final result is $\partial_\lambda F^{\lambda\rho} = 0$ which is just four of the Maxwell equations (the other four are automatically satisfied by virtue of writing the field tensor F in terms of the potential A_μ).

Let us formulate the theory canonically. Computing the canonical momentum of the spatial part of the vector potential we find that,

$$\frac{\delta L}{\delta \dot{A}_a} = E^a \tag{4.19}$$

that is, the electric field. The canonical pairs of variables are therefore $A_a(x), E^a(x)$, one pair per each point of space. The Poisson bracket of the variables is therefore given by,

$$\{A_b(x), E^a(y)\} = \delta^a_b \delta^3(x - y). \tag{4.20}$$

and notice how the Poisson bracket is zero unless we are talking about the same vector component, and the spatial points of the two variables are the same. So one sees that the variable x is acting as a "continuous index" that labels the variables at different points of space. And that different points of space involve variables that have vanishing Poisson brackets with each other. This is represented by the Dirac delta that appears on the right-hand side. The Poisson bracket for fields has the same definition as the one in mechanics, but the derivatives are functional derivatives and there is an integral on the spatial variable.

At this point it is a good idea to recall the concept of *density* we introduced in Chapter 3. If we were in a curved manifold, the Lagrangian would be given by a volume integral and its integrand must therefore be a scalar density. When one takes the functional derivative the integral is gone and the result is also a density. So strictly speaking we should have written $\tilde{E}^a$ for the electric field to denote it as a density. This is not usually done when one works in flat space, but it is good to keep in mind. The Poisson bracket of $\tilde{E}^a$ with A_b (the vector potential is not a density) gives as a result a density. As we discussed, the Dirac delta is actually a density because one can readily integrate it. The usual convention, however, is not to write it with a tilde on it in spite of its character.

The Hamiltonian is constructed via the Legendre transform, just like in mechanical systems, except for a spatial integral (which is like "summing over the continuous index"),

$$H \equiv \int d^3x \left(\tilde{E}^a(x)\dot{A}_a(x) - \tilde{\mathcal{L}} \right) \tag{4.21}$$

where we have a sum on the repeated index. As usual in Legendre transforms one uses the definition of canonical momentum to transform the Hamiltonian into an expression involving purely the canonically conjugate pairs. The result is,

$$H = \int \left(\frac{1}{2} \left[E^a(x)E^b(x)\delta_{ab} + B^a(x)B^b(x)\delta_{ab} \right] - A_0 \partial_a E^a \right) d^3x \tag{4.22}$$

where $B^a = \epsilon^{abc} F_{bc}/2$ is the magnetic field that should be viewed as a function of spatial derivatives of A_a. Notice the presence of A_0. If we work out the equation of motion of π^0 we simply get,

$$\dot{\pi}^0 = \{A_0, H\} = \partial_a E^a. \tag{4.23}$$

But recall that π^0 was zero (at any time). Therefore its time evolution must vanish too. Therefore we conclude that $\partial_a E^a = 0$, that is Gauss' law in vacuum. This equation is a constraint since it implies that the canonical variable E^a cannot be any vector one wishes but has to be divergence-less. We can now use our knowledge that constraints are generators of symmetries. Here one has to be a bit careful again since we are dealing with a field theory and therefore expressions are point dependent. In particular it is good to introduce the idea of a *smeared* constraint,

$$G(\lambda) \equiv \int d^3 x \, \lambda \partial_a E^a \tag{4.24}$$

where λ is an arbitrary function of x that is smooth, differentiable, and such that the integral on the right-hand side is well defined (one has to exercise some care if one is working in an unbounded domain, for example, but we will ignore such subtleties in this book). The smeared constraint $G(\lambda)$ is now a single number. Requiring that the smeared constraint vanish for all λ is of course equivalent to requiring that the constraint vanish at all points on the manifold. This may seem a bit superfluous, but using smearing makes calculations more straightforward and better defined. Remember that we are encountering Dirac deltas in various places, which are not functions but distributions. The latter are more cleanly handled under an integral sign. That is what the smearing provides.

If we now consider the Poisson bracket of the smeared constraint with the Hamiltonian, one indeed finds that it vanishes, showing that the "orbits" generated by the constraint leave the theory invariant. It is interesting to explore what such orbits are. To do so, let us compute them,

$$\{G(\lambda), E^a\} = 0, \tag{4.25}$$

$$\{G(\lambda), A_a\} = \partial_a \lambda, \tag{4.26}$$

so along the orbits the electric field is unchanged and the vector potential changes by the gradient of a function. But we know in Maxwell theory that the vector potential is defined up to the gradient of a function.

This is called *gauge invariance* and coincides with what we discussed in Chapter 2. And we see it emerging in the canonical theory as a consequence of the presence of Gauss' law. In this context Gauss' law is therefore called the *generator* of gauge transformations.

As we have previously discussed, the situation we encounter here is fairly generic: if you discover in your Lagrangian one of the variables you thought was a dynamical variable is behaving like a Lagrange multiplier (its canonically conjugate momentum vanishes) you have a constraint. The constraint will generate symmetries of your theory known as *gauge symmetries* and the solutions of the equations of motion of the theory will contain arbitrary parameters in them.

As we mentioned briefly before, the topic of studying theories with constraints is much larger than that we are covering here. In particular constraints can come in a variety of types, some of which are more complicated than the simple example we discuss here. It can happen, for example, that once you find your constraint it does not have a vanishing Poisson bracket with the Hamiltonian (it is not automatically preserved in time) and one has to impose its preservation, which in turn is a new constraint (which must also be preserved in time). The constraints we will consider in this book are the ones that are technically known as *first class*. Other types of constraints have to be treated with other techniques.

Incidentally, we have another constraint in the theory $\pi^0 = 0$. Does it generate a symmetry? To check that, we smear it, $\pi(\mu) \equiv \int d^3x \, \mu\pi^0$, and we note that

$$\{\pi(\mu), A_0\} = \mu \tag{4.27}$$

that is A_0 is defined up to a function of x, so we can rescale it arbitrarily, as one expects in a variable that is not physical (it is a Lagrange multiplier). The orbit of the constraint $\pi(\mu)$ does not generate any other motions in the other variables.

Finally we recover the rest of the Maxwell equations by studying the time evolution of A_a and E^a

$$\dot{A}_a = \{A_a, H\} = E_a + \partial_a A_0, \tag{4.28}$$

$$\dot{E}^a = \{E^a, H\} = \epsilon^{abc}\partial_b B_c. \tag{4.29}$$

where indices are raised and lowered using the Kronecker delta. Notice that the evolution, as anticipated, is not unique, it depends on the choice of the Lagrange multiplier A_0 and its implication is that the vector potential is defined up to the gradient of a function, which is the gauge symmetry of Maxwell theory. That is, starting from some initial data one can end in different final configurations that just correspond to different choices of gauge. The physics in terms of the electric and magnetic fields is unique, its description in terms of a vector potential is not. General relativity has the same feature, starting from some initial three metric q^{ab} and its conjugate momenta on a spatial slice, the evolution is not unique since one has freedom in choosing coordinates as one evolves. If one fixes the coordinates, then the evolution is unique. It does, however, have additional unique features that we discuss in the next section.

4.4 Totally constrained systems

We will now turn our attention to a particular type of constrained system that may appear rather artificial on first exploration, but are very important since general relativity has a behavior of the same kind. Let us start by considering Newton's mechanics. The central equation is given by the celebrated second law,

$$\vec{F} = m\vec{a} \tag{4.30}$$

where $\vec{F}$ is the applied force, m the mass of a particle, and $\vec{a}$ its acceleration. It is well known that this law is only valid in inertial reference frames. Implicit in that statement is the idea that the clock we are using to measure times has certain properties. Suppose you wanted to study classical mechanics but you had a clock that was consistently and non-linearly late (or early) such that the time it measured, t, is a known function of the time, T, in which Newton's law is valid. Could you study mechanics with your clock? The answer is of course yes, but its fundamental law would look more cumbersome than $\vec{F} = m\vec{a}$. Nevertheless, you could write it down if you wished to, and you would reach the same conclusions that a friend of yours using a better clock would reach about where the particle is, provided you translate t into T appropriately.

How can we build a Lagrangian and Hamiltonian picture for your mechanics with your lousy clock? Start from the regular action,

$$S = \int dT L(x, \frac{dx}{dT}) = \int dt\dot{T} L(x, \frac{dx}{dt}\frac{dt}{dT}) = \int dt\dot{T} L(x, \frac{dx}{dt}\frac{1}{\dot{T}}), \tag{4.31}$$

where dot means derivative with respect to t, so for the free particle this would yield,

$$S = \int dt \frac{m}{2} \frac{\dot{x}^2}{\dot{T}}.$$ (4.32)

This action is *reparametrization invariant*. That is, if you change your time parameter $t \to f(t)$ its form remains invariant. This is unsurprising since replacing t by $f(t)$ is equivalent to replacing your lousy clock by a different lousy clock. The configuration variables of the system are therefore $x(t)$ and $T(t)$. The Lagrange equations that follow are,

$$\frac{d}{dt}\left(\frac{\dot{x}}{\dot{T}}\right) = 0 = \frac{d}{dt}\left(\frac{\dot{x}^2}{\dot{T}^2}\right),$$ (4.33)

and as we anticipated the resulting equations may be complicated, depending on how bad your clock is, and the dynamics is not determined until one picks a clock. Obviously if $T = t$ then one immediately recovers $\ddot{x} = 0$.

Let us now pass to the Hamiltonian formulation by defining the momenta canonically conjugate to x and T,

$$p_x \equiv \frac{\partial L}{\partial \dot{x}} = m\frac{\dot{x}}{\dot{T}},$$ (4.34)

$$p_T \equiv \frac{\partial L}{\partial \dot{T}} = -m\frac{\dot{x}^2}{2\dot{T}^2}.$$ (4.35)

Examining the momenta we immediately realize there is a constraint,

$$\phi(x, T) = p_T + \frac{p_x^2}{2m} = 0.$$ (4.36)

If one now carries out the Legendre transform to define the Hamiltonian,

$$H = p_T \dot{T} + p_x \dot{x} - L = p_T \dot{T} + p_x \dot{x} - \frac{m}{2}\frac{\dot{x}^2}{\dot{T}},$$ (4.37)

but substituting the momenta one realizes $H = 0$, that is, the Hamiltonian vanishes. Therefore in systems like this there is no "time evolution." There is a constraint, however, and we can therefore define a "flow" with respect to the constraint as we discussed before. That flow will be the only "dynamics" that the system will see. At this point this may sound confusing. Are we not just studying the free particle? Yes, but the introduction of arbitrary possible clocks has made all possible dynamical trajectories equivalent! Therefore the arbitrariness in the clock introduced a symmetry in the system, and the only "dynamics" we can see are products of a symmetry. This does not mean the true dynamics is arbitrary, it just has to be disentangled from the use of arbitrary clocks.

It may appear that these games with lousy clocks are a bit pointless. Who would be interested in studying mechanics with an arbitrarily bad clock? This is true. But recall that in general relativity one is faced with a remarkably similar situation: the theory is invariant under arbitrary choices of time variable. It will not therefore come as a surprise that the dynamics of general relativity also has a vanishing Hamiltonian and the "evolution" is generated by constraints. Let us examine the resulting equations of motion. When one has constraints, as we discussed, it is customary to define the *total Hamiltonian* of the system as the Hamiltonian plus the constraints times Lagrange multipliers. The total Hamiltonian generates all possible flows of the system, be them symmetries or time evolution (if there is one). If the Hamiltonian vanishes, then the total Hamiltonian is a linear combination of the constraints of the systems times Lagrange multipliers. Since our system only has one constraint, we have $H_{\text{Tot}} = N(t)\phi(x, p_x, T, p_T)$. The function $N(t)$ is the Lagrange multiplier and is usually called the *lapse*. The evolution equations resulting from the Hamiltonian are,

$$\dot{x} = \{x, H_{\text{Tot}}\} = N\frac{p_x}{m}, \qquad \dot{p}_x = 0, \qquad (4.38)$$

$$\dot{T} = \{T, H_{\text{Tot}}\} = N, \qquad \dot{p}_T = 0, \qquad (4.39)$$

and we see that if N = constant one recovers usual Newtonian dynamics.

At this point it is natural to still be rather confused. How could it be that the dynamics of the system is just a symmetry? Is there not some "true" underlying dynamics in Newton's theory? The answer is yes, but it is *generated by the total Hamiltonian*. Suppose a friend of yours shows up with a different lousy clock t'. Can your friend and you make physical predictions that agree? The answer is yes. The type of questions you can ask are what is the value of x when T equals 3:30 pm? You see, 3:30 pm in terms of the variable T may correspond to 7:45 am in the parameter t or 9:38 pm in the parameter t'. But the question "what is the value of x at a given time T" is a well defined reparametrization invariant question. Such kinds of questions are called *relational* and are extremely natural questions to consider in the context of gravity. Suppose you are considering the universe, as we will do later when we consider cosmology. The universe is *everything*. Therefore, whatever clock we construct to study, its evolution is a portion of the universe. So the types of questions we will be able to ask are "what is some portion of the universe doing when the portion we use to define our clock measures the year 1776?"

Further reading

For more details on the Hamiltonian formulation of systems with constraints an excellent reference is the paper by Hanson, Regge, and Teitelboim (1976). Dirac's (2001) original treatment is also quite readable. Another reference is the book by Henneaux and Teitelboim (1992) which includes parametrized systems, a subject that is also discussed in the book by Kiefer (2006). There is also a good set of lectures by Date (2010).

Problems

1. Show that the Poisson brackets satisfy the Jacobi identity, that is, if f, g, h are three functions of phase space then,

$$\{\{f, g\}, h\} + \{\{g, h\}, f\} + \{\{h, f\}, g\} = 0. \qquad (4.40)$$

   Use it to show that if one has two constants of the motion, one can generate a (potentially) new constant of motion by considering their Poisson bracket.
2. Derive equations (4.19), (4.22), (4.28), (4.29).
3. Show how the electric field of a point particle can be described in a gauge where $\phi = 0$ (Weyl gauge).
4. The Lagrangian for a relativistic particle is given by $L = -m\sqrt{u^\mu u_\mu}$ where $u^\mu = dx^\mu/d\tau$ is the four-velocity. Notice that the action is just given by the integral of the proper time. Work out the Lagrange equations of motion and show that the Hamiltonian vanishes. Why should one have expected the Hamiltonian to vanish?
5. The action for a scalar field is given by

$$S = -\frac{1}{2} \int d^4x \left(\partial_\mu \phi \partial^\mu \phi + m^2 \phi^2 \right). \qquad (4.41)$$

   Work out Hamilton's equations of motion for it.

6. Show that the action,

$$S = \int d\tau \left(p_0 \dot{q}^0 + p\dot{q} + N(t) \left(\frac{p^2}{2m} + p_0 \right) \right), \qquad (4.42)$$

describes the motion of a non-relativistic particle in an arbitrary time parametrization. Show that for $N = 1$ one recovers ordinary Newtonian mechanics.

5

Yang–Mills theories

5.1 Kinematical arena and dynamics

Yang–Mills theories are a generalization of Maxwell's electromagnetism, and have been very successful in describing the strong and weak interactions of elementary particles. General relativity written in terms of Ashtekar variables resembles a Yang–Mills theory, hence it is appropriate to discuss some of the aspects of these theories here. To build a Yang–Mills theory one starts with a vector potential, just like in Maxwell theory, but the vector potential, instead of being a collection of four numbers, is a collection of four matrices[1]. The matrices are usually taken to be members of an algebra. An algebra is a vector space that is endowed with a multiplication operation among its members satisfying some natural properties. Like in any vector space one can introduce a basis for the algebra. An example of an algebra is the one generated by the basis of the Pauli matrices σ^i,

$$\sigma^1 = \begin{pmatrix} 0 & 1 \\ 1 & 0 \end{pmatrix}, \quad \sigma^2 = \begin{pmatrix} 0 & -i \\ i & 0 \end{pmatrix}, \quad \sigma^3 = \begin{pmatrix} 1 & 0 \\ 0 & -1 \end{pmatrix}. \tag{5.1}$$

Such algebra is called $su(2)$ algebra, because if one takes the exponential of the matrices one has an $SU(2)$ group (the exponential of a matrix is defined by using the Taylor series expansion of the exponential function). The "U" in the name coming from the fact that the exponential of the matrices are unitary matrices of two dimensions and the "S" comes from "special" which in this context means their exponentials have determinant equal to one. The Pauli matrices themselves are Hermitian. Yang–Mills theories based on $su(2)$ are useful for describing the weak interactions

---

[1]More generally, the vector potential, the electric and magnetic fields take values in a Lie algebra, and the holonomy we will discuss later takes values in a Lie group, but we will not use that terminology nor use the properties of Lie groups in this book for simplicity.

and as we mentioned above, gravity in terms of Ashtekar variables is based on an $su(2)$ vector potential as well. The strong interactions are described by QCD, quantum chromodynamics, and this is based on $su(3)$ matrices.

The vector potential will take values on the algebra, so we denote this by using boldface $\mathbf{A}_\mu$, and, just like any other element of the algebra, one can rewrite it as a linear combination of the elements of the basis of the algebra,

$$\mathbf{A}_\mu = \sum_{i=1}^{3} A_\mu^i \sigma^i, \tag{5.2}$$

where the quantities A_μ^i are the components of the vector potential in the basis given. Using the vector potential one introduces a covariant derivative,

$$D_\mu = \partial_\mu - ig/2 \; \sigma^i A_\mu^i. \tag{5.3}$$

In this context covariant does not refer to coordinate transformations in space-time but internal transformations in the group. We will not pursue the details of such transformations in our discussion. If you want, just accept D_μ as the definition of an operator. The constant g is usually called the "coupling constant" for reasons we will discuss below. The matrices that form the basis of any algebra satisfy commutation relations. If the basis of the algebra are called T^i, the commutation relations read,

$$\left[T^j, T^k\right] = i \sum_l f^{jkl} T^l \tag{5.4}$$

where the sum runs over all elements of the basis and the quantities f^{jkl} are known as structure constants of the algebra. The latter entirely determine the algebra, so if someone gives you f^{ijk} it is tantamount to giving you a complete characterization of the algebra. For the case of $su(2)$, the Pauli matrices satisfy,

$$\left[\sigma^i, \sigma^j\right] = 2i\epsilon^{ijk}\sigma^k. \tag{5.5}$$

and we are assuming from now on that repeated indices are summed from one to three. So the structure constants of the $su(2)$ algebra are just the three-dimensional Levi–Civita symbols $f^{ijk} = 2\epsilon^{ijk}$.

Just like we did with the space-time covariant derivative, one can compute the commutator of two Yang–Mills covariant derivatives,

$$[D_\mu, D_\nu] = -g/2 \, F^i_{\mu\nu} \sigma^i \tag{5.6}$$

and working out the computation explicitly one finds that $F^i_{\mu\nu}$ is given by,

$$F^i_{\mu\nu} = \partial_\mu A^i_\nu - \partial_\nu A^i_\mu + g\epsilon^{ijk} A^j_\mu A^k_\nu. \tag{5.7}$$

And in this expression we realize the similarity with the field tensor of Maxwell theory in the first two terms. The emergence of the Maxwell field tensor as a "curvature" (the commutator of two covariant derivatives) is a beautiful geometric construction to which we cannot do justice in the confines of this book since it would require dwelling on the details of Lie groups and algebras. We hope this hint of beauty will motivate you to investigate further such topics elsewhere. A good introduction is in Baez and Muniain (1994). A significant difference with the Maxwell case is that the field tensor is non-linear in the vector potential. If one views each component i of the vector potential as "a photon" it means that the "photons" are interacting with each other since the equations of motion will mix the potentials together through the non-linear term. The term photon is really inappropriate in this context (real photons do not interact with each other directly), so perhaps we should replace it with *bosons* or the more technical term *vector bosons*. One has that in Yang–Mills theories the vector bosons interact with each other. The strength of the interaction is governed by the parameter g which is therefore called the coupling constant.

This concludes the kinematics of Yang–Mills theories. But what about the dynamics? It turns out that the dynamics is formally exactly the same as in Maxwell theory, the divergence (in this case covariant divergence of the field tensor is zero,

$$D_\mu \mathbf{F}^{\mu\nu} = 0, \tag{5.8}$$

and so is its curl,

$$\epsilon^{\mu\nu\rho\sigma} D_\nu \mathbf{F}_{\rho\sigma} = 0. \tag{5.9}$$

Incidentally, it is obvious that Maxwell theory is a particular case of Yang–Mills theory in which the group is Abelian (the matrices commute with each other) and the matrices have only one component, what is called a $U(1)$ theory. If the group is Abelian and one has matrices with more than one component, one is simply dealing with a collection of Maxwell theories that do not interact with each other.

Yang–Mills theories admit a Lagrangian and Hamiltonian formulation that have many points in contact with that of Maxwell's theory we discussed a few chapters ago. The Lagrangian is given by

$$L = \frac{1}{4} \int d^3x F^i_{\mu\nu} F^{\mu\nu i} \tag{5.10}$$

and, as in the case of Maxwell theory, A^i_0 operates like a Lagrange multiplier. The canonical variables are A^i_a and E^a_i, the latter defined as minus the $0i$ component of $F^i_{\mu\nu}$. There is a constraint given by Gauss' law, which in this case reads,

$$D_a \mathbf{E}^a = 0. \tag{5.11}$$

Notice that the configuration degrees of freedom are nine in an $su(2)$ Yang–Mills theory as opposed to three in Maxwell theory but there are now three constraints instead of one. So the physical number of degrees of freedom are six, three times those of Maxwell theory, and as we mentioned these degrees of freedom interact with each other.

In Maxwell theory we learned that the vector potential is defined up to the addition of a gradient of a function. We also learned that such invariance, called gauge invariance, was generated in the canonical framework by Gauss' law. What is the analogous invariance in Yang–Mills theory? We can answer the question by considering the Poisson bracket of the smeared Gauss' law with a vector potential. Here we have to smear Gauss' law with three functions since it has an internal index, $G(\lambda) = \int d^3x \lambda^i (D_a E^a)^i$. Carrying out the calculation one gets,

$$\left\{ G(\lambda), A^i_a \right\} = \partial_a \lambda^i + g\epsilon^{ijk} A^j_a \lambda^k = (D_a \lambda)^i. \tag{5.12}$$

So we see that the notion of "gauge transformation" is more complicated than in Maxwell theory. Another difference arises if one considers the Poisson bracket of Gauss' law with the field tensor. Unlike in the Maxwell case, the Poisson bracket is non-vanishing, the field tensor transforms under gauge transformations in Yang–Mills theories. That in particular means that the electric and magnetic fields are not observable quantities in Yang–Mills theories since they are gauge dependent. We will have to search for other quantities that are invariant under gauge transformations to construct physical observables for the theory. An example of such a quantity is called *holonomy.*

5.2 Holonomies

To introduce the concept of holonomy, let us start first with Maxwell theory. Suppose we consider the circulation of the vector potential along a curve $\int_C \vec{A} \cdot \vec{ds}$. If the curve is a closed curve, due to Stokes' theorem we have,

$$\int_C \vec{A} \cdot \vec{ds} = \int_S \vec{\nabla} \times \vec{A} \cdot \vec{n}\, d^2 x, \tag{5.13}$$

where S is any surface bounded by the curve C and n is the unit normal to the surface at the point of integration. Although we will not offer a mathematical proof here, it is intuitively clear that if one were to specify the circulation *for all possible curves* on a manifold, it would be tantamount to specifying the curl of $\vec{A}$ on the manifold (if you want the curl at a given point, just take an infinitesimal curve surrounding that point, then the integral on the right is just the curl times the infinitesimal area). The reason the curl of the vector potential is important is of course because it is proportional to the field tensor, $\epsilon^{abc} F_{bc} = 2(\vec{\nabla} \times \vec{A})^a$, the right member meaning the a component of the curl. So we are saying that if you specify the circulation of the vector potential for all possible curves, you are implicitly specifying the fields.

What about Yang–Mills theory? If we tried to apply the above results directly one runs into trouble. It is obvious that the field tensor in Yang–Mills theory is something more than just the curl of a vector, since it contains the non-linear interactions among the vector potential's components. Moreover, in the Maxwell case if one performs a gauge transformation, the vector potential changes by the gradient of a function. Such a gradient does not contribute to the circulation along a closed curve. The integral of the gradient of a function along the curve simply yields the values of the functions at the endpoints, and since they are the same point in a close curve, the integral vanishes. The result is therefore gauge invariant. In the Yang–Mills case the gauge transformations are more complicated and the circulation of the vector potential will not be gauge invariant. It is clear we need a more elaborate notion to play the role of circulation of the vector potential. Such a notion is called *holonomy.*

Let us go back to the idea of parallel transport. Say you take a quantity like E_i^a and you want to carry it around a curve $\gamma^a(t)$ in space "as parallel to itself as possible," just like we did when considering curved geometries. That would mean that its covariant derivative along the curve should be zero (in this case, since we are in flat space but we are dealing with objects with Yang-Mills indices, it is just the Yang–Mills covariant derivative),

$$\dot{\gamma}^a(t)\mathbf{D}_a\mathbf{E}^b = 0, \tag{5.14}$$

where $\dot{\gamma}^a(t) = d\gamma^a(t)/dt$ is the tangent vector to the curve. This leads to the equation

$$\dot{\gamma}^a(t)\partial_a\mathbf{E}^b(t) = -ig\dot{\gamma}^a(t)\mathbf{A}_a(t)\mathbf{E}^b(t). \tag{5.15}$$

where $\mathbf{A}_a$ and $\mathbf{E}^b$ are evaluated at the relevant point of the curve $\gamma^a(t)$. One can formally integrate this equation to get,

$$\mathbf{E}^b(t) = \mathbf{E}^b(0) - ig\int_0^t ds\dot{\gamma}^a(s)\mathbf{A}_a(s)\mathbf{E}^b(s), \tag{5.16}$$

and by "formally" we mean that this is not really a solution since the variable we are solving for appears on both sides of the equation. We can, however, proceed iteratively, by inserting the left-hand side in the right-hand side. For instance, doing it once yields,

$$\mathbf{E}^b(t) = \mathbf{E}^b(0) - ig\int_0^t ds\dot{\gamma}^a(s)\mathbf{A}_a(s)\mathbf{E}^b(0) - g^2\int_0^t ds\dot{\gamma}^a(s)\mathbf{A}_a(s)$$
$$\int_0^s dw\dot{\gamma}^a(w)\mathbf{A}_a(w)\mathbf{E}^b(w), \tag{5.17}$$

and doing it twice,

$$\mathbf{E}^b(t) = \mathbf{E}^b(0) - ig\int_0^t ds\dot{\gamma}^a(s)\mathbf{A}_a(s)\mathbf{E}^b(0)$$
$$-g^2\int_0^t ds\dot{\gamma}^a(s)\mathbf{A}_a(s)\int_0^s dw\dot{\gamma}^a(w)\mathbf{A}_a(w)\mathbf{E}^b(0) \tag{5.18}$$
$$+ig^3\int_0^t ds\dot{\gamma}^a(s)\mathbf{A}_a(s)\int_0^s dw\dot{\gamma}^a(w)\mathbf{A}_a(w)\int_0^w du\dot{\gamma}^a(u)\mathbf{A}_a(u)\mathbf{E}^b(u).$$

This does not seem to be helping. However if we repeat the process indefinitely we get the sum

$$\mathbf{E}^b(t) = \sum_{n=0}^{\infty} \left((-ig)^n \int_{t_1 \geq \cdots \geq t_n \geq 0} \gamma^{a_1}(t_1)\mathbf{A}_{a_1}(t_1) \cdots \gamma^{a_n}(t_n)\mathbf{A}_{a_n}(t_n)dt_1 \cdots dt_n \right) \mathbf{E}^b(0).$$

(5.19)

A remarkable thing is that the sum actually converges for smooth finite vector potentials. The quantity in brackets, including the sum, is called the *parallel propagator*, since it is an operator that takes you from 0 to t "as parallel as possible." Given a connection[2] and a curve, the parallel propagator provides the unique solution to (5.14). If $\gamma^a(t)$ coincides with $\gamma^a(0)$ so one is propagating along a closed curve, then such a parallel propagator is called a *holonomy*[3]. Notice that the parallel propagator is a matrix. If one takes the trace of it one gets a scalar. Such a scalar happens to be invariant under gauge transformations. So it is a good candidate for an observable quantity for the Yang–Mills theory. We will actually see that it is much more than just an observable quantity.

We also see that the holonomy has some common elements with the concept of circulation. To make this connection more transparent it is good to introduce some notation that is very common in physics circles (holonomies are also heavily studied by mathematicians, who in general do not use this type of notation).

The idea is to introduce the concept of *path ordered product*,

$$P\left(\mathbf{A}_{a_1}(t_1) \cdots \mathbf{A}_{a_n}((t_n))\right),$$

(5.20)

which is just the product with the factors permuted in such a way that larger values of t_i appear first. So for instance, if $t_1 > t_2 \ldots t_n$ then

$$P\left(\mathbf{A}_{a_1}(t_1) \cdots \mathbf{A}_{a_n}(t_n)\right) = \mathbf{A}_{a_1}(t_1) \cdots \mathbf{A}_{a_n}(t_n),$$

(5.21)

---

[2]We mentioned that there was a deep link between Yang–Mills theories and geometry. It turns out that one can think of the vector potential of Maxwell and Yang–Mills theories as a connection. We will use that terminology from now on.

[3]Usage in physics is sometimes quite loose with people calling "holonomy" any parallel propagator, not just those for closed paths. We may indulge in such loose notation later in this book.

but if $t_2 > t_1 > t_3 \ldots t_n$ then

$$P\left(\mathbf{A}_{a_1}(t_1) \cdots \mathbf{A}_{a_n}(t_n)\right) = \mathbf{A}_{a_2}(t_2)\mathbf{A}_{a_1}(t_1) \cdots \mathbf{A}_{a_n}(t_n), \qquad (5.22)$$

and so on. With this notation, the integral we were considering can be rewritten as,

$$\int_{t_1 \geq \cdots \geq t_n \geq 0} \gamma^{a_1}(t_1)\mathbf{A}_{a_1}(t_1) \cdots \gamma^{a_n}(t_n)\mathbf{A}_{a_n}(t_n)dt_1 \cdots dt_n = \qquad (5.23)$$

$$\frac{1}{n!}\int_0^t P\left(\dot\gamma^{a_1}\mathbf{A}_{a_1}(t_1) \cdots \dot\gamma^{a_n}\mathbf{A}_{a_n}(t_n)\right) dt_1 \cdots dt_n = \frac{1}{n!}P\left(\int_0^t \dot\gamma(t)^a\mathbf{A}_a(t)dt\right)^n.$$

$$(5.24)$$

To check the result we suggest you work it out explicitly for two, three, etc. To get the $1/n!$ it is good to visualize the domain of integration as a hypercube and note the volume being integrated on. With the shorthand introduced in the last identity, we can now proceed to define the *path ordered exponential*,

$$P\left[\exp\left(-ig\int_0^t \dot\gamma^a(s)\mathbf{A}_a(s)ds\right)\right] \equiv \sum_{n=0}^{\infty} \frac{(-ig)^n}{n!}P\left(\int_0^t \dot\gamma(t)^a\mathbf{A}_a(t)\right)^n.$$

$$(5.25)$$

Notice that if the vector potentials commute, as in the Maxwell case, then the path ordering has no effect and the exponential collapses to the usual exponential. So this quantity becomes the exponential of the circulation in the Abelian case. We have therefore succeeded in providing a generalization of the concept of circulation to the Yang–Mills case. What about Stokes' theorem? Can we relate the path ordered exponential to an integral on the surface contained by the loop? The answer is yes, but the result, called the *non-Abelian Stokes' theorem* is quite complicated and not easy to use. The path ordered exponential of a vector potential along a closed loop is called the *holonomy* and is a matrix.

An important result, which we will offer without proof but that builds on our intuition on the circulation in the Abelian case, is called *Giles' theorem* (Giles 1981). This theorem states that if you know the trace of the holonomy along all possible loops in a manifold for a given vector potential you can reconstruct from their values all the gauge invariant information present in the vector potential. This is why we claimed that the trace of the holonomy is much more than an example of an observable for Yang–Mills theory. It is a basis for all possible observables that are functions of the connection only! This key result will be the basis for the loop representation for gauge theories and gravity, which we will discuss in detail in Chapter 8.

Further reading

There are many books dealing with Yang–Mills theories, but most of them require going through a lot of additional material related to particle physics. For a treatment that is reasonably accessible without additional topics we recommend the article by Abers and Lee (1973), and the books by Huang (1992) and Gambini and Pullin (1996), and for holonomies the book by Baez and Muniain (1994).

Problems

1. Prove equations (5.6, 5.7, 5.9).
2. Prove the Jacobi identity for the covariant derivative of Yang–Mills theory.
3. Compute the Poisson bracket of Gauss' law with itself (you may want to smear it to do the calculation). What is it proportional to?
4. Derive the formula for the infinitesimal gauge transformation of the connection and the electric field using their Poisson brackets with a smeared Gauss law. Show that the field tensor is not gauge invariant.
5. Show that the trace of the holonomy is gauge invariant. Use an infinitesimal gauge transformation. You will notice that each contribution from the non-Abelian portion of the gauge transformation cancels that of its neighbor in the product.

6

Quantum mechanics and elements of quantum field theory

6.1 Quantization

Towards the beginning of the twentieth century it became increasingly clear that physics at the atomic scale did not seem to be ruled by classical mechanics. Particles appeared in certain circumstances to behave like waves, exhibiting interference and diffraction phenomena, and outcomes of events appeared probabilistic. Eventually a new description of nature was constructed that differed from classical mechanics. In this description reality (for instance a particle) is represented by a state vector that can be encoded in a complex wavefunction $\Psi(x)$. The square of the wavefunction at a point x represents the probability that the particle be at point x. Probabilities should be normalized, so $\int dx |\Psi(x)|^2 = 1$. Wavefunctions are defined up to a phase factor $e^{i\alpha}$ with α a constant since it will not change the probabilities. If one has two particles, represented by wavefunctions $\Psi_1(x)$ and $\Psi_2(x)$, the combined probability is obtained through superposing both wavefunctions $a\Psi_1(x) + b\Psi_2(x)$. This is called the superposition principle. When one determines the joint probability for both particles one has to square the superposition and that gives rise to interference phenomena.

Physically observable quantities are represented by operators $\hat{O}$ that act on the wavefunctions. One obtains predictions for the values of the physical quantities by computing the expectation value of the operator of interest $\langle \hat{O} \rangle = \int dx \Psi^*(x) \hat{O} \Psi(x)$. Since these expectation values represent observable physical quantities they must be real numbers. This is guaranteed if observables are *self-adjoint* operators[1], that is $\hat{O} = \hat{O}^\dagger$ where $\hat{O}^\dagger$ is defined as,

---

[1] The condition for self-adjointness, in the case of infinite dimensional operators, also requires that the space of functions on which O is well defined ("domain") be the same as that of $O^\dagger$.

$$\int dx \varphi^* \hat{O}^\dagger \psi \equiv \int dx \left(\hat{O} \varphi \right)^* \psi, \tag{6.1}$$

for all possible φ and ψ. The operator $\hat{O}^\dagger$ is called the Hermitian conjugate of $\hat{O}$.

Given a classical theory one can build a quantum version of it via a procedure called *canonical quantization*. In this process one starts by formulating the classical theory in Hamiltonian fashion. One therefore has the theory formulated in terms of canonical variables q, p. One then chooses a set of physical quantities that is large enough to describe the physics of interest and promotes them to self-adjoint operators acting on a Hilbert space[2]. A Hilbert space is a space of functions endowed with an inner product. Inner products between functions usually involve an integration over the independent variable. And the result should be finite. This usually imposes restrictions over the types of functions one can consider. For instance functions that are square integrable (that is a function $f(x)$ such that $\int dx |f(x)|^2$ is finite) form a Hilbert space. The canonical variables should be promoted to operators on the Hilbert space in such a way that the Poisson brackets of the variables get promoted to commutators. So, first we therefore have variables going to operators, that is $q \to \hat{q}$ and $p \to \hat{p}$. And then that $\{q, p\} \to [\hat{q}, \hat{p}] = i\hbar \widehat{\{q, p\}} = i\hbar$. Another way to put it is that classical expressions must now hold as identities among operators (to first order in $\hbar$) if one places hats on the classical variables and replaces Poisson brackets with commutators. When one promotes classical expressions to operators there are ambiguities in how to proceed, since quantities that classically commute with each other may not commute as quantum operators. These are called factor ordering ambiguities and it is for this reason that classical expressions hold as quantum operator relations only to first order in $\hbar$. For instance the Hamilton equations become,

$$i\hbar \dot{\hat{q}} = \left[\hat{q}, \hat{H} \right], \tag{6.2}$$

$$i\hbar \dot{\hat{p}} = \left[\hat{p}, \hat{H} \right], \tag{6.3}$$

---

[2]There are restrictions on what set of physical quantities can be promoted, and in general making different choices leads to inequivalent theories, but we will not concern ourselves with these issues here. This is known in the literature as the Groenewold–van Hove (1946, 1951) theorem.

and the Poisson bracket between the canonical variables becomes $[\hat{q}, \hat{p}] = i\hbar$. The above equations are known as Heisenberg equations. Viewed from the Hamiltonian picture, quantization is a very natural procedure. If one has constraints among the canonical variables they have to be promoted to operators and they must annihilate the wavefunctions. Wavefunctions that are annihilated by the constraints are sometimes referred to as physical states. In constrained theories one sometimes considers other spaces of states that are not annihilated by the constraints, called kinematical Hilbert spaces, perhaps as a means to construct the physical space of states or to study algebras of constraints.

There is more than one picture of quantum mechanics. The one we have constructed up to now is the *Heisenberg* representation in which operators are time dependent and states are not. An alternative representation is the *Schrödinger* representation in which operators are time independent and time dependence occurs in the states. Both yield equivalent physical results, though certain calculations may be more convenient or natural in one representation than the other.

One way to construct more explicitly the Hilbert space one is interested in is to, in the language of geometric quantization, "pick a polarization" in which one considers the states to be functions of the configuration variable $\Psi(q)$. As stated before, square integrable functions of q would form a Hilbert space one could use. On this space the operator $\hat{q}$ is simply multiplicative $\hat{q}\Psi(q) = q\Psi(q)$ and the momentum is a derivative $\hat{p}\Psi(q) = -i\hbar\partial\Psi(q)/\partial q$. In certain circumstances it may be more convenient to pick a polarization in which wavefunctions are functions of the momentum. If there are many canonical variables, many possibilities arise. In the case of systems with a finite number of degrees of freedom there exists a theorem called the *Stone–Von Neumann* theorem that states that the resulting theories are all equivalent[3].

Let us consider as an example the quantization of a system that we will see plays a crucial role in the quantization of most field theories: the simple harmonic oscillator. We start with a Hamiltonian $H = p^2/(2m) + kq^2/2$ with k the spring constant. It is good to introduce the frequency $\omega^2 = k/m$ and to redefine variables a bit $q = \sqrt{\hbar/(m\omega)}Q$ and

---

[3]When the system has constraints the Stone–Von Neumann theory may not apply even in finite dimensional situations.

$p = \sqrt{m\hbar\omega}\,P$. Then the Hamiltonian simply reads $H = \hbar\omega(P^2 + Q^2)/2$. With this we have the commutation relations $[\hat{Q}, \hat{P}] = i$.

One can treat the oscillator, for instance, in the Q representation in which $\Psi(Q)$ are the wavefunctions and the eigenvalue problem for the Hamiltonian would read

$$\frac{1}{2}\left[-\frac{d^2}{dQ^2} + Q^2\right] u_\epsilon(Q) = \epsilon u_\epsilon(Q) \tag{6.4}$$

with $u_\epsilon(Q)$ the eigenvector with eigenvalue ϵ. We will not take this route here, although for this example it can be readily done, the eigenstates being given by Hermite polynomials times a Gaussian. We will instead follow a method, introduced by Dirac, that constructs the eigenvectors of the Hamiltonian by applying an operator on a starting eigenvector corresponding to the lowest eigenvalue (the vacuum). We start by defining two operators,

$$\hat{a} = \frac{\sqrt{2}}{2}(\hat{Q} + i\hat{P}), \tag{6.5}$$

$$\hat{a}^\dagger = \frac{\sqrt{2}}{2}(\hat{Q} - i\hat{P}). \tag{6.6}$$

The two operators are Hermitian conjugate of each other. The commutation relation of the operators is $\left[\hat{a}, \hat{a}^\dagger\right] = 1$. One can write the Hamiltonian in terms of these operators as,

$$\hat{H} = \frac{\hbar\omega}{2}\left(\hat{a}\hat{a}^\dagger + \hat{a}^\dagger\hat{a}\right). \tag{6.7}$$

If we now define the operator $\hat{N} = \hat{a}^\dagger\hat{a}$, we can rewrite the Hamiltonian as,

$$\hat{H} = \left(\hat{N} + \frac{1}{2}\right)\hbar\omega. \tag{6.8}$$

Let us explore a bit the properties of the operator $\hat{N}$. Consider its eigenvalues n and eigenvectors $|n\rangle$ $\hat{N}|n\rangle = n|n\rangle$. Now let us consider the norm of the states $\hat{a}|n\rangle$

$$\langle n|\hat{a}^\dagger\hat{a}|n\rangle = \langle n|\hat{N}|n\rangle = n\langle n|n\rangle \tag{6.9}$$

where we used the commutator of $\hat{a}$ with $\hat{a}^\dagger$. Since the norm of a state is by definition positive, we have that the eigenvalues of $\hat{N}$ are positive or zero. We also see that if $n = 0$ then $a|n\rangle$ is the zero vector. It should also be noted that $\hat{a}|n\rangle$ and $\hat{a}^\dagger|n\rangle$ are eigenvectors of $\hat{N}$,

$$\hat{N}\hat{a}|n\rangle = \hat{a}\left(\hat{N} - 1\right)|n\rangle = (n-1)\hat{a}|n\rangle \tag{6.10}$$

$$\hat{N}\hat{a}^\dagger|n\rangle = \hat{a}^\dagger\left(\hat{N} + 1\right)|n\rangle = (n+1)\hat{a}^\dagger|n\rangle \tag{6.11}$$

witheigenvalues $(n - 1)$ and $(n + 1)$ respectively.

We can now construct the set of eigenvectors of $\hat{N}$ obtained by acting successively on $|n\rangle$ by $\hat{a}$,

$$\hat{a}|n\rangle, \quad \hat{a}^2|n\rangle, \ldots, \hat{a}^p|n\rangle. \tag{6.12}$$

This set has a finite number of elements since the eigenvalues of $\hat{N}$ have a lower bound of zero. To put it differently, the vectors of this set vanish from a certain rank $p + 1$ on: the action of $\hat{a}$ on the non-zero eigenvector $\hat{a}^p|n\rangle$ belonging to the eigenvalue $n - p$ yields zero. This therefore requires that $n = p$. Since p is an integer number, so must be n. One can also apply $\hat{a}^\dagger$ on $|n\rangle$, which would correspond to an eigenstate of $\hat{N}$ with eigenvalue $n + 1$, apply $\left(\hat{a}^\dagger\right)^2$, which would yield an eigenstate with eigenvalue $n + 2$, and so on. In conclusion, the eigenvalues of $\hat{N}$ are the set of non-negative integers. The operator $\hat{N}$ is called the *number* operator and the operators $\hat{a}$ and $\hat{a}^\dagger$ are called the *lowering* and *raising* operators respectively (sometimes called *annihilation* and *creation* operators). The eigenstates constructed do not have unit norm. This can be easily fixed by rescaling them times a constant. We therefore define from now on,

$$|n\rangle \equiv \frac{1}{\sqrt{n!}} \left(\hat{a}^\dagger\right)^n |0\rangle, \tag{6.13}$$

which has the property that $\hat{N}|n\rangle = n|n\rangle$ but in addition to that $\langle n|n'\rangle = \delta_{n,n'}$.

The eigenvectors introduced form a basis for a quantum representation called the *number representation*. In particular, in this representation the Hamiltonian $\hat{H} = \left(\hat{N} + 1/2\right)\hbar\omega$ is automatically diagonal. The eigenvalues of the Hamiltonian are $\hbar\omega/2$, $3\hbar\omega/2\ldots$ $(n + 1/2)\hbar\omega$.

We will see that this way of handling quantum systems becomes very useful at the time of treating field theories quantum mechanically.

6.2 Elements of quantum field theory

As we argued in the section on Hamiltonian mechanics, fields are really dynamical systems with "continuous indices" accounting for the fact that variables take different values at different points in space. The same sort of attitude can be adopted when quantizing fields. We will, however, outline some developments that are commonly used when considering the quantization of fields.

Let us start by considering a scalar field (also known as a Klein–Gordon field), which is the simplest type of field theory. An electromagnetic field has many characteristics in common with a scalar field but one has to deal with the fact that the main variable has an index and the theory has gauge freedom. That makes expressions more complicated to follow and introduces some subtleties, so it is best to start with the scalar field. The Lagrangian for such a field is $L = -\int d^3x \left(\partial_\mu \varphi \partial^\mu \varphi + V(\varphi) \right)/2$. The term $V(\varphi)$ represents a potential that governs how the field interacts with itself. The field is said to have a mass if the potential contains a term $m^2\varphi^2/2$ where m is the mass of the field. The equation of motion of the Lagrangian is the wave equation with a potential term $\partial_\mu \partial^\mu \varphi - dV(\varphi)/d\varphi = 0$. If $V(\varphi) = 0$ the waves propagate freely at the speed of light. This case is called a massless free scalar field. To study the canonical formulation we define the momentum canonically conjugate to the field $\pi = \delta L/\delta \dot\varphi = \dot\varphi$. The Hamiltonian then becomes, in the simple case of the potential being the mass,

$$H = \frac{1}{2} \int d^3x \left[\pi^2 + (\nabla\varphi)^2, + m^2\varphi^2 \right] \tag{6.14}$$

where the middle term is the usual gradient of the scalar field squared.

Just like in the case of quantization of mechanical systems we will promote the field and its canonical momentum to quantum operators,

$$\varphi \to \hat\varphi, \qquad \pi \to \hat\pi \tag{6.15}$$

except that here there will be an operator per point in space. The canonical Poisson brackets become commutators,

$$[\hat{\varphi}(\vec{x}), \hat{\pi}(\vec{y})] = i\hbar\delta^3(\vec{x} - \vec{y}), \tag{6.16}$$

$$[\hat{\varphi}(\vec{x}), \hat{\varphi}(\vec{y})] = [\hat{\pi}(\vec{x}), \hat{\pi}(\vec{y})] = 0. \tag{6.17}$$

It is illuminating to expand the field in Fourier space,

$$\hat{\varphi}(\vec{x}, t) = \int \frac{d^3p}{\sqrt{(2\pi)^3}} e^{i\vec{p}\cdot\vec{x}} \hat{\varphi}(\vec{p}, t), \tag{6.18}$$

then the equation for the field, known as the Klein–Gordon equation becomes,

$$\left(\frac{\partial^2}{\partial t^2} + \left[\vec{p}^2 + m^2\right]\right)\varphi(\vec{p}, t) = 0, \tag{6.19}$$

which we identify as the equation for a harmonic oscillator of frequency $\omega(\vec{p}) = \sqrt{\vec{p}^2 + m^2}$ for each value of $\vec{p}$. So we see that in momentum space the field operates like a collection of uncoupled harmonic oscillators, one per each value of the momentum norm. Just like for the harmonic oscillator we can introduce creation and annihilation operators, there will just be one of each per value of the momentum,

$$\hat{\varphi}(\vec{p}) = \frac{1}{\sqrt{2\omega(\vec{p})}} \left(\hat{a}_{\vec{p}} + \hat{a}^\dagger_{-\vec{p}}\right), \tag{6.20}$$

$$\hat{\pi}(\vec{p}) = -i\sqrt{\frac{\omega(\vec{p})}{2}} \left(\hat{a}_{\vec{p}} - \hat{a}^\dagger_{-\vec{p}}\right), \tag{6.21}$$

and if you are puzzled about the $\omega(\vec{p})$ factors, recall that in our treatment of the Harmonic oscillator we rescaled the fields to eliminate the frequency. The presence of the $-\vec{p}$ in the creation operators is in order for the resulting fields to be self-adjoint. The commutation relations for the creation and annihilation operators that follow are,

$$\left[\hat{a}_{\vec{p}}, \hat{a}^\dagger_{\vec{p}'}\right] = \delta^3(\vec{p} - \vec{p}'). \tag{6.22}$$

We can now rewrite the Hamiltonian as,

$$\hat{H} = \int d^3p\, \omega(\vec{p}) \left\{\hat{a}^\dagger_{\vec{p}}\hat{a}_{\vec{p}} + \frac{1}{2}\left[\hat{a}_{\vec{p}}, \hat{a}^\dagger_{\vec{p}}\right]\right\}. \tag{6.23}$$

The last term is the commutator for two equal values of $\vec{p}$ and therefore it is proportional to $\delta^3(0)$; it diverges. It just represents the sum over all the momenta of the ground state energy of each harmonic oscillator. Therefore, although disquieting, it is a term that one would expect. Experiments with fields (without involving gravity) only care about energy differences. Therefore this divergent constant energy cannot be detected experimentally. The situation deteriorates significantly when one involves gravity. Then one cannot ignore this divergent term. We will not go into the details here, but if one computes the resulting stress energy tensor for the quantum field, one discovers that it has the form of a cosmological constant. The observed cosmological constant is very small, therefore the fact that, at least as a rough guess, quantum field theory suggests it is infinite unless one introduces a cutoff in momentum, is clearly problematic. This is considered one of the outstanding problems that a quantum theory of gravity will eventually have to deal with (for an alternative viewpoint see Bianchi and Rovelli (2010)).

Just as in the case of the harmonic oscillator one can introduce a basis of states labeled by the number operator, except that now one would have one such basis per value of the momentum.

At this point it is worthwhile considering the meaning of the action of a field on the vacuum $\hat{\varphi}(x)|0\rangle$. The definition of a vacuum in quantum field theory is a bit more subtle than in the quantum mechanical case. As we mentioned, in finite dimensional systems one has the Stone–Von Neumann theorem that guarantees all quantum representations (theories) are equivalent. This is not true anymore in quantum field theory and one can find many non-equivalent representations. When one is quantizing a field in flat space-time as we are in this section, it is customary to pick a representation in which the vacuum is Poincaré invariant[4]. In such a state factor orderings are chosen such that the expectation value of the linear momentum of the field and of the Hamiltonian vanish.

In such a vacuum one has that $\hat{a}_{\vec{p}}|0\rangle = 0$. The fact that we used Poincaré invariance to select the vacuum suggests that one may face issues when one studies quantum field theories in curved space-time, since in a generic curved space-time one will not have the analogue of Poincaré invariance as a guidance. This turns out to be one of the major problems in applying quantum field theories in the context of gravity. We will return to this point in Section 10.1.

---

[4]Invariant under Lorentz transformations and space-time translations.

The above quantization has been implicitly carried out in the Schrödinger picture, the operators corresponding to the fields and momenta are time independent and time dependence is present in the states. It is better to go to the Heisenberg picture to better appreciate how the quantized fields operate in space-time. Let us return to analyzing the action of the field on the vacuum. Given that the latter is annihilated by $\hat{a}_{\vec{p}}$ and recalling (6.20) and transforming in Fourier back to position space, one has that,

$$\hat{\varphi}(x)|0\rangle = \int \frac{d^3 p}{(2\pi)^3} \frac{1}{2E_{\vec{p}}} e^{-i\vec{p}\cdot\vec{x}} |\vec{p}\rangle, \tag{6.24}$$

where the energy is $E_{\vec{p}} = \sqrt{\vec{p}^{\,2} + m^2} = \omega(\vec{p})$ and $|\vec{p}\rangle \equiv \sqrt{2E_{\vec{p}}}\, \hat{a}_{\vec{p}}^{\dagger}|0\rangle$. So we see that the above expression is a linear superposition of single particle states that have well defined momentum $\vec{p}$. Except for the presence of the factor involving the energy this is the familiar non-relativistic expression for the eigenstate of position $|\vec{x}\rangle$. So we will interpret this as follows: the action of a field $\hat{\varphi}(\vec{x})$ creates a particle at position $\vec{x}$. This can be further confirmed by computing,

$$\langle 0|\hat{\varphi}(\vec{x})|\vec{p}\rangle = e^{i\vec{p}\cdot\vec{x}} \tag{6.25}$$

which allows us to reinterpret $\langle 0|\hat{\varphi}(x)$ as the position space representation of the single particle wavefunction at a point x, and the inner product as with the state $|\vec{p}\rangle$, as the single particle function of the state $|\vec{p}\rangle$, just as in ordinary quantum mechanics $\langle \vec{x}|\vec{p}\rangle \sim e^{i\vec{p}\cdot\vec{x}}$ is the wavefunction of the state $|\vec{p}\rangle$.

We now move on to consider an important concept known as the *propagator*[5]. The question we want to ask is: suppose you prepare a one particle state at $\vec{x}$, what is the probability you will find the particle at $\vec{y}$? Taking into account that the relation between the annihilation (or creation) operators in the Schrödinger and Heisenberg picture is given by $e^{iHt} a_{\vec{p}} e^{-iHt} = a_{\vec{p}} e^{-iE_{\vec{p}}t}$, the probability is given by the quantity,

$$\langle 0|\hat{\varphi}(y)\hat{\varphi}(x)|0\rangle = \int \frac{d^3 p\, d^3 p'}{(2\pi)^3} \frac{1}{4E_{\vec{p}} E_{\vec{p}'}} \langle 0|a_{\vec{p}} a_{\vec{p}'}^{\dagger}|0\rangle e^{-ip\cdot x + ip'\cdot y}$$

---

[5]Not to be confused with the parallel propagator we discussed in Chapter 5.

$$= \int \frac{d^3p}{(2\pi)^3} \frac{1}{2E_p} e^{ip\cdot(x-y)} \equiv D(x-y), \qquad (6.26)$$

and the quantity $D(x-y)$ is called the propagator. At this point the reader may be a bit puzzled about the lack of appearance of covariance of the construction. After all, we are considering a Klein–Gordon field, whose equation is Lorentz invariant. Although the above expression is apparently not Lorentz invariant, in fact it is, and this can be seen by noting that,

$$\int \frac{d^3p}{2E_{\vec{p}}} = \int d^4p \, \delta(p^2+m^2)\Theta(p^0), \qquad (6.27)$$

where $\Theta(p^0) = 1$ if $p^0 > 0$ and zero otherwise. Lorentz transformations that do not reverse the flow of time have $\Theta(p^0) = 1$ and the above expression is Lorentz invariant.

We will later need the concept of a *time ordered product* of two fields, and this leads to the concept of the *Feynman propagator*,

$$\Delta_F(x-y) = \langle 0|T\hat{\varphi}(x)\hat{\varphi}(y)|0\rangle = \begin{cases} D(x-y) & x^0 > y^0 \\ D(y-x) & y^0 > x^0 \end{cases} \qquad (6.28)$$

Notice that the Feynman propagator depends on the space-time points x and y, not just the spatial portions, and one can actually write a covariant expression for it in terms of an integral,

$$D_F(x-y) = \int \frac{d^4p}{(2\pi)^4} \frac{-i}{p^2+m^2-i\epsilon} e^{-ip\cdot(x-y)}, \qquad (6.29)$$

where ϵ is a small positive number. The origin of the ϵ is that one really has to evaluate the integral with $\epsilon = 0$ but it has poles at $p^0 = \pm E_{\vec{p}}$ and should be evaluated using a contour in the complex plane that goes above the pole at $E_{\vec{p}}$ and below the pole at $-E_{\vec{p}}$. The addition of the $i\epsilon$ accomplishes this while treating the integral as along the real line. Although we will not derive this expression, it should be noted that it can be understood as the Fourier transform of the Green's function of the Klein–Gordon equation with the chosen boundary conditions. In momentum space the Klein–Gordon equation is $(p^2+m^2)\,\varphi(p) = 0$, and in effect the expression (6.29) is Fourier transforming the inverse of that expression. We finish this section with some jargon: the propagator is sometimes referred to as the two-point correlation function or two-point Green's function.

6.3   Interacting quantum field theories and divergences

Up to now we have considered the quantization of free field theories. In that case, particle states are eigenstates of the Hamiltonian. There is no interaction and no scattering among the particles. In order to have interactions one needs non-linear terms in the Hamiltonian that are different from the mass term. We will consider only one example of interacting theory, the one resulting from adding in the Hamiltonian of a scalar field a term $\lambda \phi^4/4!$. These theories are known as "lambda phi fourth." In this context λ is called the "coupling constant" and is a dimensionless number that we will assume to be small.

The first thing we would like to try to compute is the two-point correlation function in the interacting theory,

$$\langle \Omega | T \hat{\varphi}(x) \hat{\varphi}(y) | \Omega \rangle \tag{6.30}$$

where we denote by Ω the fact that the vacuum of the interacting theory is not necessarily going to be the same as in the free theory. We start by writing the Hamiltonian as,

$$H = H_0 + H_{\text{int}} = H_0 + \int d^3 x \frac{\lambda}{4!} \hat{\varphi}^4(\vec{x}). \tag{6.31}$$

The Hamiltonian enters into (6.30) in two places: the definition of the vacuum and the definition of the field at a time t, since the evolution of the field in terms of the Hamiltonian is given by

$$\hat{\varphi}(t, \vec{x}) = \exp(i\hat{H}t)\hat{\varphi}(0, \vec{x})\exp(-i\hat{H}t) \tag{6.32}$$

in the Heisenberg representation. We would like to obtain expressions in powers of the coupling constant λ.

We start with the field $\hat{\varphi}(x)$. At any time t_0 we can expand it in terms of creation and annihilation operators,

$$\hat{\varphi}(t_0, \vec{x}) = \int \frac{d^3 p}{(2\pi)^3} \frac{1}{\sqrt{2E_{\vec{p}}}} \left(a_{\vec{p}} e^{i\vec{p}\cdot\vec{x}} + a_{\vec{p}}^\dagger e^{-i\vec{p}\cdot\vec{x}} \right), \tag{6.33}$$

and to get the field at any time we could use (6.32). If λ is small most of the evolution will be due to H_0, so it is useful to define a field in the *interaction picture* using (6.32) but with H_0. One therefore gets,

$$\hat{\varphi}_I(x) = \int \frac{d^3p}{(2\pi)^3} \frac{1}{\sqrt{2E_{\vec{p}}}} \left(a_{\vec{p}} e^{-ip\cdot x} + a_{\vec{p}}^{\dagger} e^{ip\cdot x} \right) |_{x^0 = t-t_0}. \tag{6.34}$$

The challenge now is to express the fully interacting field in terms of $\hat{\varphi}_I$. To do it, we start by noting that one can trivially write,

$$\hat{\varphi}(x) = \hat{U}^{\dagger}(t, t_0) \hat{\varphi}_I(x) \hat{U}(t, t_0) \tag{6.35}$$

with $\hat{U}(t, t_0) = \exp\left(i\hat{H}_0(t - t_0)\right) \exp\left(-i\hat{H}(t - t_0)\right)$, the time evolution operator in the interaction picture. We need to find this operator in terms of $\hat{\varphi}_I$ since we know the latter explicitly. If one takes the time derivative of the time evolution operator one gets,

$$i\frac{\partial \hat{U}(t, t_0)}{\partial t} = e^{i\hat{H}_0(t-t_0)} \left(\hat{H} - \hat{H}_0\right) e^{-i\hat{H}(t-t_0)} = \hat{H}_I \hat{U}(t, t_0), \tag{6.36}$$

where $H_I = \exp(i\hat{H}_0(t - t_0))\hat{H}_{\text{int}} \exp(-i\hat{H}_0(t - t_0)) = \int d^3x \lambda \hat{\varphi}_I^4/(4!)$. To integrate this equation we will use the same technique we used in the previous chapter to deal with equation (5.15). The end result, instead of being a path-ordered exponential is a time-ordered exponential,

$$U(t, t_0) = T\left(\exp\left[-i\int_{t_0}^{t} dt' \hat{H}_I(t')\right]\right) \tag{6.37}$$

$$\equiv 1 + (-i)\int_{t_0}^{t} dt_1 \hat{H}_I(t_1) + \frac{(-i)^2}{2!}\int_{t_0}^{t} dt_1 \int_{t_0}^{t} dt_2 T\left(\hat{H}_I(t_1)\hat{H}_I(t_2)\right) + \dots$$

and this is known as Dyson's formula.

We now should deal with the fact that the vacuum in the interacting theory is really different from the vacuum in the free theory. For reasons of brevity we will not discuss this point in detail. The reader who wants further details can consult the treatment in Peskin and Schroeder (1995), on which we have based this section, for example. But remarkably, the expectation value taken in the vacuum of the interactive theory can be related to that taken in the free theory by,

$$\langle \Omega | T\left(\phi(x)\phi(y)\right) | \Omega \rangle = \lim_{T \to \infty} \frac{\langle 0 | T\left(\phi_I(x)\phi_I(y) \exp\left[-i\int_{-T}^{T} dt H_I(t)\right]\right) | 0 \rangle}{\langle 0 | T\left(\exp\left[-i\int_{-T}^{T} dt H_I(t)\right]\right) | 0 \rangle}, \tag{6.38}$$

where T is supposed to have a small imaginary component $T \to \infty$ $(1 - i\epsilon)$.

So the computation of correlation functions has now been boiled down to computing expressions of the form,

$$\langle 0|T\left(\hat{\varphi}_I(x_1)\hat{\varphi}_I(x_2)\cdots\hat{\varphi}_I(x_n)\right)|0\rangle. \tag{6.39}$$

To handle these expressions it is usual to follow a procedure that breaks them into products of free propagators, known as Wick's theorem. Before discussing the theorem we need to introduce two definitions involving products of operators. The first definition is that of normal ordering. Whenever one has a product of creation and annihilation operators, it is said to be normal-ordered if the creation operators appear to the left of the annihilation operators. For instance, let us consider our field and separate it into positive and negative frequency[6] parts: $\hat{\varphi}_I(x) = \hat{\varphi}_I^+(x) + \hat{\varphi}_I^-(x)$, where

$$\hat{\varphi}_I^+(x) = \int \frac{d^3p}{(2\pi)^3}\frac{1}{\sqrt{2E_{\vec{p}}}}\hat{a}_{\vec{p}}e^{ip\cdot x}, \qquad \hat{\varphi}_I^-(x) = \int \frac{d^3p}{(2\pi)^3}\frac{1}{\sqrt{2E_{\vec{p}}}}\hat{a}_{\vec{p}}^\dagger e^{-ip\cdot x}, \tag{6.40}$$

then we would call the normal ordered product of two fields,

$$N\left(\hat{\varphi}_I(x)\hat{\varphi}_I(y)\right) = \hat{\varphi}_I^+(x)\hat{\varphi}_I^+(y) + \hat{\varphi}_I^-(x)\hat{\varphi}_I^+(y) + \hat{\varphi}_I^-(y)\hat{\varphi}_I^+(x)$$
$$+ \hat{\varphi}_I^-(x)\hat{\varphi}_I^-(y). \tag{6.41}$$

The importance of normal order is that by putting the annihilation operators to the right, the normal ordered version of a product of operators immediately vanishes when acting on the vacuum.

The second definition we need is that of a contraction of two fields. A contraction of two fields is the vacuum expectation value of their time ordered product. For two fields $\hat{\varphi}(x)$ and $\hat{\varphi}(y)$ we saw that such an expectation value is given by the Feynman propagator. Let us denote the contraction of two fields as $\langle\hat{\varphi}_I(x)\hat{\varphi}_I(y)\rangle$.

Wick's theorem states that given a time ordered product of a finite number of operators, it is equal to a sum of normal ordered products from which $0, 1, 2, \ldots$ contractions have been removed. To make this more

---

[6]The terminology "positive" and "negative" frequency is understood by recalling that the energy is given by $E = i\partial/\partial t$ and positive frequencies are associated with positive energies.

concrete, let us consider an example with four fields (we drop the subscript *I* for brevity),

$$T\left(\hat{\varphi}(x_1)\hat{\varphi}(x_2)\hat{\varphi}(x_3)\hat{\varphi}(x_4)\right) = N\left(\hat{\varphi}(x_1)\hat{\varphi}(x_2)\hat{\varphi}(x_3)\hat{\varphi}(x_4)\right) \quad (6.42)$$
$$+\langle\hat{\varphi}(x_1)\hat{\varphi}(x_2)\rangle N\left(\hat{\varphi}(x_3)\hat{\varphi}(x_4)\right) + \langle\hat{\varphi}(x_1)\hat{\varphi}(x_3)\rangle N\left(\hat{\varphi}(x_2)\hat{\varphi}(x_4)\right)$$
$$+\langle\hat{\varphi}(x_1)\hat{\varphi}(x_4)\rangle N\left(\hat{\varphi}(x_2)\hat{\varphi}(x_3)\right) + \langle\hat{\varphi}(x_2)\hat{\varphi}(x_3)\rangle N\left(\hat{\varphi}(x_1)\hat{\varphi}(x_4)\right)$$
$$+\langle\hat{\varphi}(x_2)\hat{\varphi}(x_4)\rangle N\left(\hat{\varphi}(x_1)\hat{\varphi}(x_3)\right) + \langle\hat{\varphi}(x_3)\hat{\varphi}(x_4)\rangle N\left(\hat{\varphi}(x_1)\hat{\varphi}(x_2)\right)$$
$$+\langle\hat{\varphi}(x_1)\hat{\varphi}(x_2)\rangle\langle\hat{\varphi}(x_3)\hat{\varphi}(x_4)\rangle + \langle\hat{\varphi}(x_1)\hat{\varphi}(x_3)\rangle\langle\hat{\varphi}(x_2)\hat{\varphi}(x_4)\rangle$$
$$+\langle\hat{\varphi}(x_1)\hat{\varphi}(x_4)\rangle\langle\hat{\varphi}(x_2)\hat{\varphi}(x_3)\rangle.$$

So the first term is just the normal ordered product, the next six terms correspond to removing one contraction, and the last three terms to removing two contractions. The theorem is proved by complete induction. This may sound a bit daunting, but is actually very useful. As we argued, normal ordered products vanish when one takes their vacuum expectation values. That means that if one wishes to compute the vacuum expectation value of the left-hand side of (6.42) only the last three terms in the right-hand side would contribute. And recalling that the contractions are just the Feynman propagators, this implies that,

$$\langle 0|T\left(\hat{\varphi}(x_1)\hat{\varphi}(x_2)\hat{\varphi}(x_3)\hat{\varphi}(x_4)\right)|0\rangle = D_F(x_1 - x_2)D_F(x_3 - x_4)$$
$$+D_F(x_1 - x_3)D_F(x_2 - x_4)$$
$$+D_F(x_1 - x_4)D_F(x_2 - x_3). \quad (6.43)$$

We have therefore succeeded in reducing all expressions of interest in the interacting theory to products of Feynman propagators of the free theory. Recalling that the Feynman propagator $D_F(x - y)$ is the amplitude for a particle prepared at x to be detected at y leads to a graphical representation given by a line joining x to y. We can therefore pictorially represent the above identity as shown in Fig. 6.1. This type of pictorial representation in space suggests particles are created at a point, travel in space-time, and are annihilated at another point, and are called *Feynman diagrams*.

Up to now we considered a generic product of fields. But let us get back to our original purpose, to compute terms in the expression of the propagator in the interacting theory. This requires considering the product of two fields times the evolution operator that we found in Dyson's

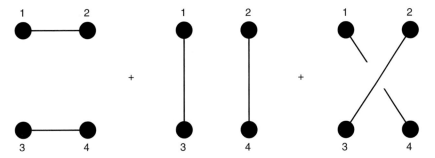

Fig. 6.1 *A pictorial representation of the right hand side of equation (6.43). These types of diagrams are known as Feynman diagrams.*

formula. Such an operator can be expanded in powers of the coupling constant. The first non-trivial contribution will therefore be given by,

$$\langle 0|T\left(\hat{\varphi}(x)\hat{\varphi}(y)(-i)\int dt\int d^3z \frac{\lambda}{4!}\hat{\varphi}^4(z)\right)|0\rangle = -\frac{3i\lambda}{4!}D_F(x-y)\int d^4z D_F(z-z)^2$$

$$-3i\lambda\int d^4z D_F(x-z)D_F(y-z)D_F(z-z). \tag{6.44}$$

And again this can be pictorially represented as shown in Fig. 6.2. The first term in Fig. 6.2 consists of a particle traveling from x to y and a pair of "virtual particles" (particles that are created out of the vacuum and annihilate each other after some time) creating and annihilating at z. The second term consists of a particle moving from x to y that spawns a virtual particle at z. The latter diagrams are known as *tadpole* diagrams. In both terms we see the appearance of $D_F(z-z) = D_F(0)$. Such terms are problematic, since they correspond to the integral,

$$D_F(0) = \int \frac{dp^4}{(2\pi)^4} \frac{-i}{p^2+m^2-i\epsilon} \tag{6.45}$$

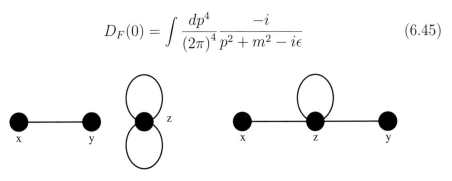

Fig. 6.2 *A pictorial representation of the right hand side of equation (6.44), which is the first non-trivial term in the evaluation of the interacting theory propagator.*

which is clearly divergent for large values of the momentum p. This is called an *ultraviolet divergence*. This brings us to the final concept we want to introduce, *renormalizability*.

6.4 Renormalizability

As we just discussed, the expansion of the propagator for the interacting theory has divergences. Actually both terms in the above expression diverge. It turns out the divergence of the first term is canceled when one does not ignore, as we incorrectly did, that the vacuum of the interacting theory is not the same as that of the free theory (that is, use in full equation (6.38)). So we will ignore that term. But even taking the correct vacuum the tadpole divergence remains. Worse, when one includes higher orders in perturbation theory, tadpoles proliferate. This is where the beauty of the Feynman diagrams comes into effect, just by inspecting the diagrams we can foretell what the various corrections are. The types of diagrams are shown in Fig. 6.3. Each straight line connecting different points will be given by a Feynman propagator D_F that will be finite. The closed loops correspond to $D_F(0)$ and will be divergent. Writing schematically, one will end up with a series of the form,

$$D_F + D_F D_F(0) D_F + D_F D_F(0) D_F D_F(0) D_F + \dots \tag{6.46}$$

$$= D_F \left(1 + D_F(0) D_F + (D_F(0) D_F)^2 + \dots \right) \tag{6.47}$$

$$= D_F \frac{1}{1 - D_F(0) D_F} = \frac{1}{p^2 + m^2} \left[\frac{1}{1 - \frac{D_F(0)}{p^2 + m^2}} \right] = \frac{1}{p^2 + m^2 - D_F(0)}. \tag{6.48}$$

We presented this resummation in a handwaving way but it can actually be carried out carefully. What we see is that the final result for the contribution of all the tadpoles to the propagator of the interacting theory is to reproduce the propagator of the free theory but with a (divergent) shift in the mass $m^2 \rightarrow m^2 - D_F(0)$. This leads to the idea of

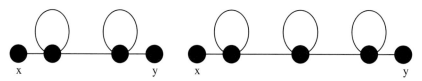

Fig. 6.3 *The tadpole diagrams that appear at higher orders in perturbation theory.*

renormalization. Perhaps we were naive in thinking that the parameter m that appears in the Lagrangian is the mass one measures when one does a particle physics experiment. That value is called the *bare* value of the parameter, because what one actually measures in the interacting theory is $m^2 - D_F(0)$, the *dressed* value of the parameter. That is, in reality the parameter that appears in the Lagrangian is divergent, and the mass one measures is finite due to the interactions of the theory.

At this point the reader may exhibit some skepticism about this construction. After all, the tadpoles are just the tip of the iceberg. One can imagine many Feynman diagrams with closed lines and those will all diverge. Where are we going to absorb those divergences now that we have redefined the only parameter in the theory, the mass? Remarkably, it turns out that for this theory everything fits together very well and indeed all divergences are canceled out by the renormalization of the mass we discussed above plus a renormalization of the coupling constant and the wavefunction.

The situation deteriorates dramatically in theories like gravity, where there is a dimensionful coupling constant. In natural units, Newton's constant has dimensions of length squared. So say one tried to study the theory perturbatively by saying that the metric is flat plus small perturbations, $g_{\mu\nu} = \eta_{\mu\nu} + h_{\mu\nu}$ where $|h_{\mu\nu}| \ll 1$. The action for the equations of general relativity we discussed in Chapter 3 is given by the integral of the space-time curvature,

$$S = \frac{1}{16\pi G} \int d^4x \sqrt{-\det g}\, R, \tag{6.49}$$

see Carroll (2003) for how to derive the Einstein equations from it. If we schematically expand the curvature in terms that involve derivatives of the metric (either second derivatives or squares of first derivatives) and terms that are quadratic in the metric, one gets something like,

$$S = \frac{1}{16\pi G} \int d^4x \left[(\partial h)^2 + (\partial h)^2 h + \ldots \right]. \tag{6.50}$$

In order to make contact with the formalism we developed for the scalar field, there is a small detail to attend to and it is the fact that the scalar field has dimensions of inverse length, whereas h here is dimensionless. To fix this we rescale the fields, $h = \sqrt{G}\,\tilde{h}$ and one gets,

$$S = \frac{1}{16\pi} \int d^4x \left[(\partial \tilde{h})^2 + \sqrt{G} (\partial \tilde{h})^2 \tilde{h} + \ldots \right], \tag{6.51}$$

so we see that the action is cast as a "free theory" plus an "interaction" term with $\sqrt{G}$ operating as a coupling constant. The free theory describes a linear symmetric massless tensor field and corresponds to particles of spin 2, known as *gravitons*. Such a theory has two degrees of freedom that represent the two "polarization" states of a gravitational wave, just like in Maxwell theory with electromagnetic waves. Since the coupling constant has dimensions of length, it means that as one goes to higher and higher orders in perturbation theory, the integrals have to have more and more factors of the momentum in the numerator in order to keeps things dimensionally correct. Each integral becomes divergent in a worse fashion and things do not fit together as for the $\lambda\varphi^4$ theory. The resulting theory in the gravitational case is non-renormalizable and therefore lacks predictive power. The only way to cancel the divergences is to add an infinite number of new terms to the bare Lagrangian. Stelle (1977) showed that if one included quadratic terms in the curvature in the action, the resulting theory can be made renormalizable. Alas, at a classical level the theories with higher powers of the curvature are already pathological in various ways, so this cannot be considered a solution to the problem. Abandoning the Stelle proposal one is left with canceling the divergent terms by hand one by one. Most of these terms are irrelevant at low energies, so one can make sense of perturbative quantum gravity as an effective theory at low energies (see for instance Donoghue (1994)). But at high energies one is out of luck. This is the reason perturbative quantum gravity runs into trouble and new ideas are needed.

The discussion we have done of this subject has been very superficial, a more detailed but still very readable discussion of perturbative quantum gravity is present in the recent review article by Woodard (2009).

One last comment has to do with a set of ideas that has gained some interest, and it is the scenario of *asymptotic safety*, first introduced by Weinberg (1979). A full discussion requires ideas of renormalization group that exceed the level of this text, but a simple example from Woodard can help shed some light on what is attempted. As we saw, in a non-renormalizable theory like general relativity one will in principle need infinitely many counterterms in the action to cancel the divergences, and that makes the theory devoid of predictive power. The idea behind asymptotic safety is that somehow the infinitely many counterterms combine in ways that allow predictions from the theory. The counterterms that appear in the action involve higher derivatives of the metric in order to keep things dimensionally uniform, since the coupling constant has dimensions of length. The example that Woodard introduces is that of a harmonic oscillator with a higher order term in the Lagrangian,

$$L = \frac{m}{2}\dot{q}^2 - \frac{m\omega^2}{2}q^2 - \frac{gm}{2\omega^2}\ddot{q}^2 \tag{6.52}$$

where g is the coupling constant. Such an action leads to equations of motion of order higher than second and one cannot determine the trajectory given $q(0)$ and $\dot{q}(0)$, implying there are new degrees of freedom in the theory. The theory therefore "does not have predictive power" given the usual set of initial data. Let us assume the constant is small and treat the extra term in the action as a perturbation, and seek a perturbative solution to the equations of motion,

$$q(t) = \sum_{n=0}^{\infty} g^n x_n(t). \tag{6.53}$$

Substituting in the equations of motion and keeping orders in powers of g one gets,

$$\ddot{x}_0 + \omega^2 x_0 = 0, \tag{6.54}$$

$$\ddot{x}_1 + \omega^2 x_1 = -\frac{1}{\omega^2}\frac{d^4 x_0}{dt^4}, \tag{6.55}$$

$$\ddot{x}_2 + \omega^2 x_2 = -\frac{1}{\omega^2}\frac{d^4 x_1}{dt^4}, \tag{6.56}$$

and similarly for higher orders. The higher derivative terms appear as sources and do not appear at zeroth order. If one is given $q_0(0)$ and $\dot{q}_0(0)$ one can integrate the system. One can actually solve the equations exactly and sum the perturbative series, to get,

$$q = q_0 \cos{(k_+ t)} + \frac{\dot{q}_0}{k_+} \sin{(k_+ t)}, \tag{6.57}$$

with $k_+ = \omega\sqrt{1 - \sqrt{1 - 4g}}/\sqrt{2g}$. Notice that the functional form of the result is the same as if one had not included the extra term. The presence of the counterterm was to shift the frequency from ω to k_+. There is no loss of predictive power.

The expectation is that a similar situation could develop in general relativity, that the infinite counterterms could somehow be absorbed in a finite redefinition of bare parameters, Newton's constant, and the cosmological constant. Progress along these lines has happened recently but there is not a general consensus on whether it provides a viable theory of quantum gravity (see Lauscher and Reuter (2002, 2005), Percacci (2006)).

Further reading

We have only covered an infinitesimal amount of the vast field of quantum field theory. The book by Peskin and Schroeder (1995) gives a very good introduction and we have based this chapter on it. Also the article by Woodard (2009) is particularly clear about the issues in perturbative quantum gravity and contains several useful analogies.

Problems

1. Show that (6.28) holds.
2. Prove Wick's theorem.
3. Derive expression (6.44).
4. Show that computing the commutator at equal times via equation (6.26) one indeed gets zero.
5. Prove equation (6.25).

7

General relativity in terms of Ashtekar's new variables

7.1 Canonical gravity

A detailed treatment of the Hamiltonian formulation of general relativity exceeds the scope of this book. Suffice it to say that it took over 50 years for it to be worked out. It is not dramatically different from the treatment we did for Maxwell theory, but the calculations are significantly more complicated. Here we will only outline some of the elements of such a formulation.

The action for general relativity is given by the integral in space-time of the scalar curvature times the square root of minus the determinant of the metric, known as the Einstein–Hilbert Lagrangian,

$$S = \frac{1}{16\pi G} \int d^4x \sqrt{-\det(g)} R. \tag{7.1}$$

If one examines carefully the expression for the curvature, one reaches the conclusion that the g^{00} component of the metric and the g^{0i} components appear without their time derivatives. They are therefore Lagrange multipliers. They are related to the *lapse* and *shift* we introduced in Section 3.6 by $g^{00} = 1/N^2$ and $g^{0i} = N^i/N^2$. The spatial components of the metric g^{ij} become the configuration variables. The canonically conjugate momenta to the metric $\tilde{\pi}_{ij}$ (since we are generically in a curved space-time we will use tildes to denote densities), are related to the tensor known as the *extrinsic curvature* that we introduced as characterizing how the spatial surface is curved with respect to space-time.

The Lagrange multipliers are, as in the Maxwell case, associated with constraints. In fact, just as we anticipated in the example of the parametrized particle, the total Hamiltonian for the theory is just a combination of the constraints. The constraints associated with the shift form a vector. The "flow" that they generate is associated with spatial diffeomorphisms.

That is, they represent the fact that general relativity is invariant under spatial coordinate transformations. The constraint associated with the lapse (usually called Hamiltonian or superHamiltonian constraint) represents the invariance of the theory under the choice of any possible deformation of the spatial surface. It is akin to the reparametrization invariance of time we discussed in the mechanical example, except that because one can reparametrize time at each point in space, the end result is a deformation of the spatial three surface when viewed within space-time. The theory has therefore six configuration degrees of freedom (g^{ij} is a symmetric 3×3 matrix) and four constraints, which leaves us with two degrees of freedom, just like Maxwell theory.

7.2   Ashtekar's variables: classical theory

Ashtekar (1986) introduced a new set of variables to describe general relativity in the canonical language. One half of Ashtekar's variables are the densitized triads we discussed in the chapter on general relativity, $\tilde{E}_i^a$. The other half behave like an $SU(2)$ Yang–Mills connection A_a^i and constitute the configuration variables, the densitized triads being their canonically conjugated momentum,

$$\left\{ A_a^i(x), \tilde{E}_j^b(y) \right\} = 8\pi G \beta \delta_b^a \delta_j^i \delta^3(x - y). \tag{7.2}$$

with G Newton's constant. The quantity β is a constant known as the *Barbero–Immirzi parameter*. This constant can in principle take any non-zero (even complex) value and implies that there really is a one-parameter family of Ashtekar variables. The resulting classical theories are all the same, one is just using different canonical coordinates to describe them. They are related by what in the canonical language is called a *canonical transformation* of variables.

What do these variables have to do with gravity? The densitized triads can be used to reconstruct the spatial metric via

$$\tilde{\tilde{q}}^{ab} = \det(q)q^{ab} = \tilde{E}_i^a \tilde{E}_j^b \delta^{ij}, \tag{7.3}$$

as we discussed in Chapter 3. The connection can be used to reconstruct the extrinsic curvature, which, as we mentioned, is a measure of how the three metric evolves in space-time. The relation is given by

$$A_a^i = \Gamma_a^i + \beta K_a^i, \tag{7.4}$$

where $\Gamma^i_a = \Gamma_{ajk}\epsilon^{jki}$ is the spin connection and $K^i_a = K_{ab}\tilde{E}^{ai}/\sqrt{\det(q)}$.

In terms of the above variables, the Lagrangian for general relativity becomes,

$$L = \frac{1}{8\pi G\beta}\int d^3x\left(\tilde{E}^a_i\dot{A}^i_a + \underset{\sim}{N}\,\epsilon_{ijk}\tilde{E}^a_i\tilde{E}^b_j F^k_{ab} + N^a\tilde{E}^b_i F^i_{ab} + \lambda^i\left(D_a\tilde{E}^a\right)^i\right).$$

(7.5)

To get to this Lagrangian from (7.1) is a laborious process and we will not need it in what follows so we will not discuss it in detail here. In particular we have absorbed in the definition of the Lagrange multipliers a factor of the square root of the determinant of the metric in N, turning it into a density of weight -1, and[1] a factor of β. Also we have chosen $\beta = i$, we will discuss this point in more detail later.

The action has been massaged to a form in which it is obvious that $\tilde{E}^a_i$ and A^i_a are canonically conjugate and the lapse $\underset{\sim}{N}$, the shift N^a, and the gauge parameter λ^i are Lagrange multipliers. The theory has seven constraints associated with them. Notice the counting of degrees of freedom, one has nine configuration degrees of freedom in the A^i_a's and seven constraints, which leaves general relativity as a theory with two degrees of freedom, just like Maxwell theory.

The first set of constraints are just a Gauss' law,

$$\mathcal{G}^i = D_a\tilde{E}^a_i = 0,$$

(7.6)

the next set are the so-called *momentum constraint* or *vector constraint*

$$V_b = \tilde{E}^a_i F^i_{ab} = 0,$$

(7.7)

and finally one has the *Hamiltonian constraint*,

$$H = \epsilon_{ijk}\tilde{E}^a_i\tilde{E}^b_j F^k_{ab} = 0.$$

(7.8)

[1]An agreement about normalizations does not seem to exist in the literature and some papers are not very explicit in their choices. Some people choose to include a factor $8\pi G\beta$ in the definition of the densitized triad equation (3.35). This makes the fundamental Poisson bracket (7.2) look cleaner, being just proportional to deltas. However factors of $8\pi G\beta$ crop up in various other equations. Most authors seem to simply ignore the factor (or set $8\pi G = 1$), something that may make things confusing at the time of coupling to matter. By reabsorbing the factor of β our constraints will look more like those of papers that ignore the factor.

The total Hamiltonian of the theory is a linear combination of these constraints. One can study their orbits. Gauss' law generates $su(2)$ gauge transformations in the variables, just like in the Yang–Mills case we discussed in a previous chapter. The momentum constraint generates orbits that are related to spatial diffeomorphisms. We say "are related" because they also mix in a bit of a gauge transformation as well. If you want to purely generate diffeomorphisms you need to consider a linear combination of the momentum constraint with the Gauss' law, called the *diffeomorphism constraint*.

$$C_a = V_a - A_a^i \left(D_b \tilde{E}_i^b \right) \tag{7.9}$$

Which of course you are free to do, since one can linearly combine constraints. To see that this constraint is related to diffeomorphims one can smear, as we have done before with other constraints, in this case with a test vector field $\vec{N}$, $C(\vec{N}) \equiv \int d^3x N^a C_a$. If one computes the Poisson bracket of a function of the canonical coordinates $f(\tilde{E}, A)$ with the diffeomorphism constraint one gets,

$$\left\{ C(\vec{N}), f(\tilde{E}, A) \right\} \sim \mathcal{L}_{\vec{N}} f, \tag{7.10}$$

that is, the "orbit" generated by the constraint in phase space is just similar to the Lie derivative along N^a (the proportionality factor depends on the choice of normalization.

The Hamiltonian constraint generates the "time evolution" in terms of the zeroth component of the x coordinate. Since the action is invariant under reparametrizations of that (and all the other) coordinate such "time evolution" is not real. General relativity is a totally constrained system as the ones we discussed at the end of Chapter 4. The total Hamiltonian of general relativity is a combination of constraints times Lagrange multipliers, that is,

$$H_T = \int d^3x \left\{ \underset{\sim}{N} \epsilon_{ijk} \tilde{E}_i^a \tilde{E}_j^b F_{ab}^k + N^a \tilde{E}_i^b F_{ab}^i + \lambda^i \left(D_a \tilde{E}^a \right)^i \right\}. \tag{7.11}$$

Remarkably if you write the Hamilton equations that result from the total Hamiltonian you will indeed reproduce the usual Einstein equations. But the insight gained in the Hamiltonian treatment is that the true "dynamics" of the theory is not captured by the evolution in an arbitrary parameter x^0 (or t). One will have to work harder to construct observables for the theory. Unsurprisingly, the Hamiltonian constraint does not have a neat geometric action in terms of the canonical variables since we broke the space-time into space and time. If you smear it and consider its Poisson brackets with functions of the canonical variables one just gets complicated expressions without a clear geometric meaning, like the one we got in the diffeomorphism constraint.

Notice that the (unconstrained) phase space of general relativity written in terms of Ashtekar's new variables coincides with the unconstrained space of a Yang–Mills theory. In fact one can view general relativity as a different type of Yang–Mills theory, one which in addition to Gauss' law has four extra constraints and has a vanishing Hamiltonian. Of course these are profound differences and the dynamics of general relativity and Yang–Mills theory are vastly different. So it is clear from the outset that one will not be able to import into general relativity all of the techniques that were used to deal with Yang–Mills theory. But some will be useful.

As we mentioned, the above expression for the Hamiltonian involves making the choice of Barbero–Immirzi parameter $\beta = i$. This was the original choice of Ashtekar in 1986. This choice has the virtue of simplifying expressions. But it has the problem that the variables become complex. When one quantizes the theory one has to struggle to make sure that one is recovering real general relativity as opposed to complex general relativity, which is a dramatically different theory. More modern analyses have shown that one can live with the extra complication introduced by choosing β to be a real number. Essentially everything remains unchanged except that the Hamiltonian constraint becomes (Barbero 1995),

$$H = \epsilon_{ijk}\tilde{E}_i^a\tilde{E}_j^bF_{ab}^k + 2\frac{(\beta^2+1)}{\beta^2}\left(\tilde{E}_i^a\tilde{E}_j^b - \tilde{E}_j^a\tilde{E}_i^b\right)\left(A_a^i - \Gamma_a^i\right)\left(A_b^j - \Gamma_b^j\right) = 0.$$

$$(7.12)$$

At first the second term (which vanishes for $\beta = i$) looks intimidating: it involves the variables Γ_a^i which have a complicated non-polynomial relationship with the densitized triads and therefore appear a nightmare at the time of quantization. Modern techniques have been introduced to handle such terms successfully. We will comment briefly on how to deal with such a term when we discuss the loop representation of the Hamiltonian constraint, but largely ignore it in this book.

An aspect that one needs to consider when one has a theory with many constraints is the issue of the *constraint algebra*. Constraints need to be preserved in time. Therefore their Poisson bracket with the total Hamiltonian must vanish. Except that in the case of general relativity the total Hamiltonian is a linear combination of constraints. This implies that Poisson brackets among constraints must vanish (or at least be proportional to constraints). This makes the constraints of general relativity what are called in the Dirac terminology *first class*: when one takes Poisson brackets of the constraints one gets linear combinations of the constraints as a result. To study the algebra, in addition to smearing the diffeomorphism constraint with a test vector field $C(N)$ as above, and Gauss' law $G(\lambda)$ with a test field with internal indices λ^i, like we did in the Yang–Mills case, one needs to smear the Hamiltonian constraint with a scalar[2] function N,

$$H(N) = \int d^3x N \frac{\tilde{E}^{ai} \tilde{E}^{bj} F^k_{ab} \epsilon_{ijk}}{\sqrt{\det(q)}}, \tag{7.13}$$

where we had to divide by $\sqrt{\det(q)}$ so we ended up with a density of weight one that could be integrated on the manifold. In terms of the above smeared constraints the constraint algebra among Gauss' laws reads,

$$\{G(\lambda), G(\mu)\} = G([\lambda, \mu]) \tag{7.14}$$

where $[\lambda, \mu]^i \equiv \lambda_j \mu_k \epsilon^{ijk}$. And so we see that "commuting" two Gauss' laws with different smearings is equivalent to a single Gauss' law evaluated on the commutator of the smearings. The algebra among diffeomorphisms is given by,

$$\left\{ C(\vec{N}), C(\vec{M}) \right\} = C(\mathcal{L}_{\vec{N}} \vec{M}), \tag{7.15}$$

and we see that the effect of one diffeomorphism on the other is to "shift its smearing." Why is this? The smearing functions are test fields, they are not functions of the canonical variables. The diffeomorphism constraint therefore does not generate diffeomorphisms on them. It does generate diffeomorphisms on everything else. So it is equivalent to leaving everything else fixed and shifting the smearing. This immediately allows us to compute the action of a diffeomorphism on a Gauss' law,

---

[2]Here we depart from our previous choice of smearing with a density of weight -1 since this form of the constraint, with the determinant in the denominator, is the one that ends up being quantized, as we will see in the next chapter.

$$\left\{ C(\vec{N}), G(\lambda) \right\} = G(\mathcal{L}_{\vec{N}}\lambda), \tag{7.16}$$

again, it shifts everything except the test field λ^i so it is equivalent to leaving everything else fixed and shifting the test field. The Gauss' law has a vanishing Poisson bracket with the Hamiltonian constraint. The diffeomorphism constraint with a Hamiltonian gives a Hamiltonian with its smearing shifted,

$$\left\{ C(\vec{N}), H(M) \right\} = H(\mathcal{L}_{\vec{N}}M). \tag{7.17}$$

Finally, the Poisson bracket of two Hamiltonians is a diffeomorphism,

$$\{H(N), H(M)\} = C(\vec{K}) \tag{7.18}$$

where the vector $K^a = \tilde{E}_i^a \tilde{E}^{bi} \left(N \partial_b M - M \partial_b N \right) / (\det(q))$. Notice that the vector K^a is not just a combination of derivatives of test functions but actually involves the canonical variables. That means that although the Poisson bracket of two Hamiltonian constraints is proportional to a diffeomorphism, the proportionality factor depends on the canonical variables. This is unlike any of the other Poisson brackets. Since upon quantization the canonical variables get promoted to operators, to ensure the proportionality of the Poisson brackets to a combination of constraints will become problematic.

7.3 Coupling to matter

In Chapter 9 we will need to consider the theory coupled to matter, in particular to a scalar field, so let us consider how to do that. As we discussed in Chapter 6, the Lagrangian for a scalar field is $L = -\int d^3x \left(\partial_\mu \varphi \partial^\mu \varphi + V(\varphi) \right)$ in flat space-time. In curved space-time we need to make explicit the dependence on the metric and its determinant in the volume element, so the action would be,

$$S = \int d^4x \left(-g^{\mu\nu} \partial_\mu \varphi \partial_\nu \varphi - V(\varphi) \right) \sqrt{-\det(g)}. \tag{7.19}$$

If we define the canonical momentum of the scalar field as usual $\tilde{\pi} = \delta L / \delta \dot{\varphi}$, the Hamiltonian can be rewritten as,

$$H = \int d^3x N \left(\frac{\tilde{\pi}^2}{\sqrt{\det(q)}} + \sqrt{\det(q)} \left(q^{ab} \partial_a \varphi \partial_b \varphi + V(\varphi) \right) \right) + N^a \tilde{\pi} \partial_a \varphi,$$

$$(7.20)$$

where N and N^a are the lapse and shift. We can proceed to write the Hamiltonian in terms of Ashtekar variables straightforwardly,

$$H = \int d^3x \frac{N}{\sqrt{\det(q)}} \left(\tilde{\pi}^2 + \tilde{E}_i^a \tilde{E}^{bi} \partial_a \varphi \partial_b \varphi + \det(q) V(\varphi) \right) + N^a \tilde{\pi} \partial_a \varphi,$$

$$(7.21)$$

From there we can read the contribution of the scalar field to the Hamiltonian and diffeomorphism constraint,

$$C(\vec{N})_\varphi = \int d^3x N^a \tilde{\pi} \partial_a \varphi \qquad (7.22)$$

$$H(N)_\varphi = \int d^3x \frac{N}{\sqrt{\det(q)}} \left(\tilde{\pi}^2 + \tilde{E}_i^a \tilde{E}^{bi} \partial_a \varphi \partial_b \varphi + \det(q) V(\varphi) \right). \ (7.23)$$

and these should be added (multiplied by $8\pi G\beta$) to the diffeomorphism and Hamiltonian constraint of the gravitational field, respectively. This would couple gravity to a scalar field.

7.4 Quantization

The first new perspective introduced by the new variables is a shift in what is the natural quantum representation to use. As we mentioned in the traditional treatment with metric variables, the configuration variables were the three metric q_{ab} and its canonically conjugate momentum $\tilde{\pi}^{ab}$ was related to the extrinsic curvature. When people attempted to canonically quantize gravity with these variables the natural thing to consider were wavefunctions of the metric $\Psi(q_{ab})$. So one could potentially think of them as giving the probability of a certain geometry occurring.

With the new variables, given that the configuration variable is A_a^i, the most natural thing to consider are wavefunctions $\Psi(A_a^i)$. This is the *connection representation*. Similar representations have been used in the quantization of Maxwell and Yang–Mills theory. But notice that this is sort of the opposite of what one was doing in the metric variables case, since here the metric is associated with the triads and the wavefunctions depend on their conjugate momentum. Let us continue with this quantization. The canonical variables get promoted to operators. The connection is a multiplicative operator,

$$\hat{A}_a^i \Psi(A) = A_a^i \Psi(A), \tag{7.24}$$

and the triads are functional derivatives[3],

$$\widehat{\tilde{E}}_i^a \Psi(A) = -i \frac{\delta \Psi(A)}{\delta A_a^i}, \tag{7.25}$$

with commutation relations,

$$\left[\hat{A}_b^j(y), \widehat{\tilde{E}}_i^a(x) \right] = i \delta_b^a \delta_i^j \delta^3(x - y). \tag{7.26}$$

One now needs to promote the constraints to operator equations holding on the quantum states as we discussed in the previous chapter. This part gets really hard and is, actually, incomplete in this representation for reasons we will discuss below. Let us start with Gauss' law

$$\hat{\mathcal{G}}^i \Psi(A) = -i D_a \frac{\delta \Psi(A)}{\delta A_a^i}. \tag{7.27}$$

And here one has adopted a *factor ordering* in which we have placed the triads to the right of the connections. Remarkably, if one smears the quantum Gauss' law we have just introduced and study its action on the quantum state one notices that the action of the constraint on the quantum state is equivalent to shifting the argument of Ψ by an infinitesimal (in the sense that the parameter λ is small) gauge transformation (recall (5.12)),

$$\left[1 + \int d^3x \lambda^i(x) \hat{\mathcal{G}}^i(x) \right] \Psi(A) = \Psi(A + D\lambda) = \Psi(A), \tag{7.28}$$

and the last identity is related to the fact that the constraints annihilate the state. So this constraint, as a quantum operator, is imposing on the quantum states the same symmetry that its vanishing imposed classically: it is telling us that the functions $\Psi(A)$ cannot be arbitrary. They have to be *gauge invariant* functions of the connection.

---

[3]Here we take $8\pi G = \beta = 1$ so the equations better resemble those in the literature. Most of the literature on this subject is old, when several other conventions were used.

Let us now study the vector constraint,

$$\hat{V}_a \Psi = -i\hat{F}^i_{ab} \frac{\delta \Psi(A)}{\delta A^i_b}. \tag{7.29}$$

If one smears this constraint (more precisely the diffeomorhism constraint we discussed above) one can check that its action on the state is to shift the argument of Ψ by an infinitesimal (in the sense that $\vec{N}$ is small) diffeomorphism,

$$\left[1 + \hat{C}(\vec{N})\right] \Psi(A) = \Psi(A + \mathcal{L}_{\vec{N}} A), \tag{7.30}$$

as exhibited by the presence of the Lie derivative. So again we see that the constraint, when promoted to a quantum operator, generates in the quantum states the same symmetry it generated in the classical theory as an orbit via the Poisson bracket. It imposes that the states be functions of A that are invariant under diffeomorphisms.

What about the Hamiltonian constraint? Here we face several problems. The first is that this constraint does not generate a simple geometric action even at a classical level. Remember this constraint generates the evolution in x^0 of the Einstein equations. This is not reflected in any simple geometric action on the variables living on a spatial surface. Moreover the presence of two triads causes computational problems. Generically your quantum state $\Psi(A)$ will be a spatial integral of a gauge invariant function of A^i_a. Such a state would be annihilated by Gauss' law and the diffeomorphism constraint. If one does not integrate spatially, the diffeomorphism constraint will just shift things around and not vanish on such a state. When the first functional derivative in the Hamiltonian constraint acts on such a state, it eliminates the integral in the state. When the second one acts, it will produce a Dirac delta, since it is not acting on a functional but on a function of the coordinates and that means that at the end of the day one ends up with a Dirac delta. However, both functional derivatives associated with the triads are acting at the same point. This means the result will be $\delta^3(x - x)$ or $\delta^3(0)$ which is not a well defined quantity. Here we face one of the problems we anticipated in the introduction of this book, namely that in quantum theory operators are in general distributions and their products are not well defined. This is precisely what is happening here. In usual quantum field theory in a fixed background space-time there are techniques to deal with these problems called *regularization*. One way to handle a product of operators $\hat{O}_1(x)\hat{O}_2(x)$ is to write $\int d^3y\hat{O}_1(x)\hat{O}_2(y)f_\epsilon(x,y)$ where $f_\epsilon(x,y)$ is a function that in the limit $\epsilon \to 0$ becomes $\delta^3(x-y)$. So one works out the calculation of interest and then at the end takes the limit $\epsilon \to 0$. If the result is finite, one has "regularized" the expression of interest. Unfortunately this is not a good procedure in the case in which one does not have a fixed background geometry. The reason for this is that the functions $f_\epsilon(x,y)$ are an external fixed structure that one imposes on the theory. That breaks background independence. For instance, it will stop the Hamiltonian constraint becoming a scalar under diffeomorphisms. This spoils the internal consistency of the theory and in particular the constraint algebra.

An additional problem surrounding the quantization is that a central element in a quantum theory is the inner product. Without an inner product a quantum theory does not have physical predictive power: one can neither compute expectation values nor transition probabilities. A suitable inner product for dealing with wavefunctions of a connection that are invariant under gauge transformations and diffeomorphisms was unknown until the early 1990s. It was the introduction of loop quantum gravity techniques that allowed us to define such an inner product, which encounters its most simple and natural expression when discussed in terms of the loop representation.

One final problem surrounding the quantization of the Hamiltonian constraint is the fact that the constraint that one should be promoting to a quantum operator should have a density weight $+1$. That means that we have to consider, as we did when we smeared it, the constraint with a $\sqrt{\det(q)}$ in the denominator. The reason for this is that, in a manifold without any background geometry, there is a naturally defined density one object (the Dirac delta), but there is no naturally defined density weight two object. And one cannot multiply Dirac deltas to build such an object. Therefore there is no chance in such a setting to represent double densitized objects. The determinant in the denominator at first appears as a rather formidable problem, making the constraint highly non-polynomial in the variables. There is, however, a trick to handle the situation, first introduced by Thiemann (1996). The trick consists in noting that the Poisson bracket of the volume

$$V = \int d^3x\sqrt{\det(q)} = \frac{1}{6}\int d^3x\sqrt{|\tilde{E}_i^a\tilde{E}_j^b\tilde{E}_k^c\epsilon^{ijk}\epsilon_{abc}|} \qquad (7.31)$$

with the connection A_a^i,

$$\left\{A_c^k, V\right\} = \frac{\tilde{E}_i^a \tilde{E}_j^b \epsilon_{gbc}\epsilon^{kij}}{\sqrt{\det(q)}}.$$ (7.32)

This allows us to write the (density weight +1 version of the) Hamiltonian constraint as a polynomial,

$$H(M) = \int d^3x M \left\{A_c^k, V\right\} F_{ab}^k \tilde{\epsilon}^{abc}.$$ (7.33)

It turns out that a similar trick can actually be used to generate the second term in the Hamiltonian (the one that arises when one considers a real value for the Barbero–Immirzi parameter β). For reasons of brevity we will not discuss it here, more details can be found in Thiemann's paper and in his book. From now on we will assume that the Barbero–Immirzi parameter is real valued.

Perhaps it is good to mention at this point that for a period of time, until Thiemann introduced this technique to deal with the singly densitized Hamiltonian constraint, there were many attempts to define the doubly densitized Hamiltonian constraint via point splitting and other types of regularizations and to attempt to find solutions to the constraint. All of this activity that took place in the period between 1988 and 1996 led to further developments in the field that eventually superseded it, and it is not considered a valid part of loop quantum gravity any more. Many of these abandoned results are nevertheless playing an important role in the modern understanding in subtle ways. For instance, Jacobson and Smolin (1988) noted that the version of the Hamiltonian constraint regulated via point splitting annihilated states consisting of holonomies of the Ashtekar connection constructed along loops without intersections. This in turn inspired Rovelli and Smolin (1988, 1990) to explore the loop representation. The result is still playing a role today, as we will see in the next chapter the modern definition of the Hamiltonian constraint only acts at vertices and this is directly tied to the result of Jacobson and Smolin. Also, some of the results concerning the doubly densitized Hamiltonian are playing a role in recent work by Laddha and Varadarajan (2010) on parametrized field theories.

The fact that we did not know how to work with functions of a connection $\Psi(A)$ that are gauge and diffeomorphism invariant in a controlled mathematical fashion, with a suitable inner product, and the difficulties of promoting the Hamiltonian constraint to an operator, led to the development of an alternative representation called the loop representation, which we will treat in the next chapter.

Further reading

The book by Ashtekar (1988) presents a good account of the understanding of gravity in the connection representation and of the old way of introducing the Ashtekar variables. More modern treatments can be found in the books by Rovelli (2007) and Thiemann (2008).

Problems

1. Compute the constraint algebra between Gauss' law and the diffeomorphism constraint and between two diffeomorphisms.
2. Show that the diffeomorphism constraint actually generates diffeomorphisms on a function of the scalar field $f(\varphi)$.
3. Show that the quantum Gauss' law generates gauge transformations.
4. Prove Thiemann's identity (7.32).
5. Show that if one regulated the Hamiltonian constraint with a function $f(x,y)_\epsilon$ it would fail to be a scalar under diffeomorphims (do the calculation classically).

8

Loop representation for general relativity

The difficulties we discussed in the previous chapter with the connection representation, in particular the inability to have good control over the space of solutions to Gauss' law and diffeomorphism constraints led Rovelli and Smolin (1988) to explore a new representation for quantum gravity called the loop representation. A similar representation had been independently introduced for Maxwell and Yang–Mills theories by Gambini and Trias (1980, 1981, 1986). In this chapter we will describe the construction of the loop representation.

8.1 The loop transform and spin networks

As we argued in the Yang–Mills chapter, according to Giles' theorem (Giles 1981), if one gives the trace of the holonomy of a connection for all possible loops on a manifold one is implicitly giving all the gauge invariant information of the connection. That is, traces of holonomies constitute a basis of gauge invariant functions of a connection. In the previous chapter we saw that in the connection representation of quantum gravity the Gauss' law constraint imposed that the wavefunctions be gauge invariant functions of the connection. Therefore traces of holonomies constitute a basis of solutions for Gauss' law. This suggests that one could expand a state in such a basis,

$$\Psi[A] = \sum_{\gamma} \Psi[\gamma] W_{\gamma}[A], \tag{8.1}$$

where the "sum" is a formal sum over all possible closed loops and the coefficients are functions that depend on a loop $\Psi[\gamma]$. The traces of holonomies[1] $W_{\gamma}[A]$ are defined as,

---

[1] We have used the notation common in particle physics, the letter W referring to Kenneth Wilson, the object W_{γ} sometimes referred to as "Wilson loop." Rovelli and Smolin used the notation T^0 for the same object.

$$W_\gamma[A] = \mathrm{Tr}\left(P\left[\exp\left(-\oint_\gamma \dot\gamma^a(s)\mathbf{A}_a(s)ds\right)\right]\right), \qquad (8.2)$$

as we discussed in the Yang–Mills chapter. Just as we work with functions $\Psi[A]$ it should be possible to work with the coefficients of the expansion of such a state, that is, directly with $\Psi[\gamma]$. Doing that is working in the *loop representation*. The above expansion is known as the *loop transform*. You can see the parallels with going to the momentum representation in quantum mechanics. There one has a basis of states $\exp(ikx)$ labeled by a number (or vector) k and one expands $\Psi(x) = \int dk\Psi(k)\exp(ikx)$ and works with the coefficients of the expansion $\Psi(k)$.

Working with a space of functions of loops may sound strange at first but it has very natural properties, particularly in the context of gravity. For instance, it is easy to solve the diffeomorphism constraint in such a space. One simply has to consider functions of loops that are invariant under smooth deformations of the loops. Such functions have been studied by mathematicians for many years, they are known as knot invariants and the branch of mathematics that studies them is known as knot theory. Such functions do not have to be abstract and hard to handle. For instance, consider two curves $\gamma^a(s_1)$ and $\eta^b(s_2)$ in three-dimensional flat space. The integral

$$\mathrm{Linking}(\gamma_1, \gamma_2) = \frac{1}{4\pi}\oint_\gamma ds_1 \oint_\eta ds_2 \partial^a\left(\frac{1}{|\gamma^d(s_1) - \eta^d(s_2)|}\right)\epsilon_{abc}\dot\gamma^b(s_1)\dot\eta^c(s_2),$$
$$(8.3)$$

is such that it gives as a result one if the two curves are linked and zero otherwise as shown in Fig. 8.1. This quantity is a function of two loops rather than one, but it illustrates the idea of a function of loops that is invariant under smooth deformations of the loops. It is known as the Gauss linking number.

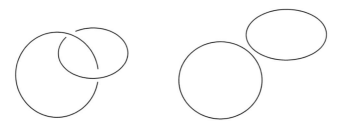

Fig. 8.1 *The Gauss linking number would be one for the loops on the left, zero for the loops on the right.*

Using the loop transform (8.1) one can translate the action of operators to the loop representation. Before we do that we have to deal with a problem: the loop basis is an over-complete basis. Imagine a three-dimensional vector space and choose a basis of vectors consisting of four vectors. First of all, the vectors cannot be independent. Moreover, if one chose to expand vectors in terms of this basis one would have four components but they would not all be independent since we know three components is all that is needed to characterize a vector in three dimensions. A very similar situation occurs here except that the space is non-linear. Not all traces of holonomies on a manifold are independent and non-linear relations exist among them. That means, in particular, that the types of functions of loops $\Psi(\gamma)$ that one can consider are severely constrained. The over-completeness comes from identities satisfied by the traces of $su(2)$ matrices. Given two such matrices $\mathbb{A}$, $\mathbb{B}$, one can check that,

$$\text{Tr}\,(\mathbb{A})\,\text{Tr}\,(\mathbb{B}) = \text{Tr}\,(\mathbb{A}\mathbb{B}) + \text{Tr}\,\left(\mathbb{A}\mathbb{B}^{-1}\right). \tag{8.4}$$

This has as immediate consequence that, given two loops γ and η, one will have,

$$W_\gamma[A]W_\eta[A] = W_{\gamma\circ\eta}[A] + W_{\gamma\circ\eta^{-1}}[A], \tag{8.5}$$

where by η^{-1} we mean the loop η traversed in the opposite direction and $\gamma \circ \eta$ means the loop obtained by going around γ and then along η. Given that the matrices are unitary one has that $W_\gamma[A] = W_{\gamma^{-1}}[A]$. Also given the cyclic property of the matrix traces one has that $W_{\gamma\circ\eta}[A] = W_{\eta\circ\gamma}[A]$. The problem lies in that these identities can be combined with each other into further identities of increasing complexity adding more loops. These identities are known as *Mandelstam identities*.

To address the problem of the over-completeness of the basis of loops we will have to talk about representations of algebras a little bit. As we stated, the connection that one uses to describe general relativity when using Ashtekar variables is a $su(2)$ connection. We mentioned that an algebra was characterized by structure constants and that in the case of $su(2)$ the structure constants were the Levi–Civita symbols,

$$\left[\sigma^i, \sigma^j\right] = 2i\epsilon^{ijk}\sigma^k. \tag{8.6}$$

The Pauli matrices satisfy the above relation. However, there are infinitely many matrices that also satisfy the above relations, they just are not 2×2 matrices but $(N+1) \times (N+1)$ matrices with $N = 1, 2, 3, \ldots$. These larger matrices are called "different representations" of the algebra. The most compact one is called the *fundamental representation* which for $su(2)$ is the one given by the Pauli matrices. It turns out one can use matrices in *any* representation to construct a connection and, with it, parallel transport operators along curves, just like we did with the Pauli matrices. The parallel transport operator along an open curve is a matrix. One can then "tie up" the indices of those matrices at intersections by contracting them with certain objects called "intertwiners." The resulting object is called "spin network." It is a graph with intersections and with "colored" lines, the "coloring" given by the number N associated with the dimensionality of the matrices of the holonomies (some authors label the lines with half-integers $N/2$). An example is shown in Fig. 8.2. The "coloring" allows us to keep track of which representation is being used in constructing the parallel transport operator along each line. To spell this all out in detail will require more knowledge of group theory and representations than the level of this course allows. Fortunately we will not need much of this in detail in what follows. Just to give a glimpse, a result in group theory is that the higher dimensional representations of $su(2)$ are in one-to-one correspondence with tensor products of the fundamental representation (tensor product in this context means just

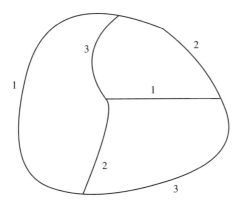

Fig. 8.2 *An example of a spin network. The numbers on the lines characterize the dimension of the parallel transport matrix along the line. A line of color N corresponds to a matrix of dimension $N + 1$.*

putting the representations side by side without any mixing of elements). So you can view a holonomy in the higher representations as a bunch of holonomies in the fundamental representation running in parallel. The intertwiners are then suitable tensor products of ϵ_{ijk} and δ^i_j where the indices i, j refer to the Pauli matrices. You can see more details in the appendices of the paper by Rovelli and Upadhya (1998).

Spin networks actually constitute a basis for all gauge invariant functions which minimize the degree of over-completeness of the loop basis (and for trivalent intersections eliminates it entirely) and therefore are a good basis in terms of which to expand a state of quantum gravity. The easiest way to see this is to introduce an inner product in the space. The construction of the inner product can be worked out in detail by introducing a formal measure in the space of connections known as the Ashtekar–Lewandowski (1997) measure. We will not pursue the details here but just note the end result. Given two spin networks s and s' (meaning graphs with an assignment of colorings and intertwiners as mentioned above), construct the spin network states ψ_s and $\psi_{s'}$. The inner product of these two states is 1 if s and s' are related to each other by a diffeomorphism (if they can be smoothly deformed into each other) and zero otherwise. Even leaving out the details of how such an inner product is arrived at, its simple nature makes it quite compelling.

Now, suppose that a given spin network state was not independent from the others. Then one could write it as a linear combination of other spin network states,

$$\psi_s = \sum_m c_m \psi_{s_m}, \qquad (8.7)$$

with s_m not diffeomorphic to s. But if one now considers the inner product of such a state with itself, on the left-hand side we would have 1 and on the right-hand side zero since the inner product of s_m with s is zero as they are not diffeomorphic to each other. Therefore it follows that spin network states are linearly independent. This result was first encountered in the context of lattice gauge theory in a rather different language by Kogut and Susskind (1975). The first to put it in the spin network language were Rovelli and Smolin (1995). Spin networks were first introduced as an abstract mathematical construction by Roger Penrose (1971) although the paper presciently suggested a possible connection with quantum gravity.

The above result is strictly true for trivalent spin networks (meaning with intersections with at most three lines converging into them). When one has intersections with valences four or higher (and as we shall see one needs such states) there is some redundancy in the choice of intertwiners and one may have to choose a basis of independent spin networks to carry out the above construction.

It is interesting to note that using the Ashtekar–Lewandowski inner product one can also make rigorous sense of the loop transform and its inverse, but the details exceed the scope of this treatment, see for instance Thiemann (1998).

To close this section it is worth emphasizing that the presentation we made of the Ashtekar–Lewandowski measure is deceptively simple. In reality it has rather dramatic consequences that will be quite influential in loop quantum gravity. To appreciate this, we need to go back to the original definition of holonomy as a measure of the curvature. If you recall, one parallel-propagated an object around a closed circuit and when one returned to the point of origin one noted an "angle shift." Such angle shift was dependent on the curvature of the region bounded by the curve. It also clearly depends on the size of such region. If one, for instance, deformed the curve slightly to encompass a different area, one would get an angle shift that changed slightly as well, at least if the connection in the intervening region is a smooth function. However, the inner product of Ashtekar and Lewandowski operates in a vastly different way. It tells us that curves that are slightly shifted from each other either yield the same result if they are diffeomorphic to each other, or yield a completely different result if they are not. What this is telling us is that such a measure is therefore leading us to work with connections that are not smooth functions, but distributions. This is not a surprise in quantum field theory since we learned that distributions arise all over the place in that context in Chapter 6. But it has implications. The first is that the connection is not going to be a well defined quantum operator, in spite of its holonomy being a well defined quantum operator. The triad will not be a well defined quantum operator either. Since it consists of a functional derivative, acting on a one-dimensional object as a holonomy it produces a result that is distributional. This is easily fixed by considering a related operator, the *triad flux* operator, obtained by integrating the triad along a two surface. We will discuss this in the next section.

A large set of mathematical details have been left out in this section to make the material readily accessible at the undergraduate level. Readers interested in more details are referred to the book by Thiemann (2008). One last point worthy of mention is that at first sight it appears that the Asthekar–Lewandowski measure is sort of arbitrary. This is far from the case, powerful mathematical results known as the "LOST-F" theorem[2] (Lewandowski, *et al.* (2006), Fleischhack (2006)) actually show that with some mild assumptions (using the holonomy based algebra and background independence) it is actually *unique*. So the kinematical underpinning of loop quantum gravity is quite strong, it is basically the only choice available, at least in a diffeomorphism invariant context.

8.2 Geometric operators

Now that we have constructed the loop representation, it would be good to study the action of some operators in it. In a theory that has a gauge symmetry, one really can only interpret physically operators that are gauge invariant. In the canonical language that means quantities that have vanishing Poisson brackets with the constraints. They are sometimes called *Dirac observables* or, in shorthand, observables. For instance in Maxwell's theory the electric field is an observable quantity whereas the electric potential is not (one can measure physically potential differences, not absolute values of the potential). In gravity this means that one should consider objects that have vanishing Poisson brackets with all constraints. This is problematic. In fact we do not know of a single such object, at least for the theory in vacuum. If one couples the theory to certain forms of matter, it is possible to use the matter as a "reference frame" and define quantities with respect to it that are Dirac observables. We do not have space to work out this construction in this book, we refer the reader to the work of Giesel, Hofmann, Thiemann, and Winkler (2007).

Ignoring that point for a while, we would like to concentrate on geometrical operators, that is, quantum operators associated with certain aspects of the geometry. Before that, let us start with the simplest operator one can define in the loop representation: the holonomy. The action of a trace of a holonomy along a spin network as a quantum operator $\hat{h}_{s'}$ on a given spin network state $|s\rangle$ is simply given by the state $|s' \cup s\rangle$ which is just

---

[2]The name is based on the initials of the authors Lewandowski, Okołów, Sahlmann, Thiemann and, independently, Fleischhack.

the spin network based on s and s' (for non-intersecting s and s'). To prove this requires a bit more knowledge of spin networks than we have introduced, but you can get a flavor of this result by acting on the loop transform (8.1) with a holonomy. On the right-hand side there will be two holonomies, which you can combine into a single holonomy along the composed loops using the Mandelstam identities.

The next such operator we would like to consider is the *area operator*. Given a certain surface Σ, we need to build an expression for its area. Let us choose coordinates such that the surface is characterized by $x^3 = 0$. Obviously the resulting operator will act on spin network states not invariant under diffeomorphisms. Then the area of the surface is given by,

$$A_\Sigma = \int_\Sigma dx^1 dx^2 \sqrt{\det q^{(2)}} \tag{8.8}$$

where $\det q^{(2)} = q_{11}q_{22} - q_{12}^2$ is the determinant of the metric in the space spanned by the coordinates x^1 and x^2. The latter can be rewritten as $\det q^{(2)} = \epsilon^{AB}\epsilon^{CD}q_{AC}q_{BC}/2$ where the indices $A \dots D$ go from 1 to 2. This can be further rewritten as $\det q^{(2)} = \epsilon^{3ab}\epsilon^{3cd}q_{ac}q_{bc}/2$ where now the indices $a \dots d$ go from 1 to 3 formally, but because of the ϵ in practice still run only from 1 to 2. We now consider the identity that gives the inverse of a matrix using the Levi–Civita symbols,

$$q^{ab} = \frac{\epsilon^{acd}\epsilon^{bef}q_{ce}q_{df}}{3!\det(q)}. \tag{8.9}$$

But in the Asthekar variables we have $\tilde{E}_i^a \tilde{E}^{bi} = \det(q)q^{ab}$. Therefore we can write $\det q^{(2)} = \tilde{E}_i^3 \tilde{E}^{3i}$ and have an expression for the area of the surface given by,

$$A_\Sigma = \int_\Sigma dx^1 dx^2 \sqrt{\tilde{E}_i^3 \tilde{E}^{3i}}. \tag{8.10}$$

We would like to promote this quantity to a quantum operator. According to the rules of canonical quantization we should promote the triads E_i^3 to quantum operators $-8i\pi G\beta\delta/\delta A_3^i$ (we are reinstating the $8\pi G\beta$ factor in the Poisson bracket so we can show explicitly that the area has the right units). Unfortunately, this cannot be straightforwardly done. On the one hand, we have a product of two such operators and we have already argued that this causes problems. Worse, there is the square root to contend with. To gain a handle on these problems we will use a technique called *smearing* of the operators. We have already encountered this technique when dealing with constraints.

We introduce a one parameter family of smearing functions $f_\epsilon(x,y)$ such that in the limit $\epsilon \to 0$ they tend to the Dirac delta in two dimensions. To make this precise, one has that,

$$\lim_{\epsilon \to 0} \int_\Sigma d^2 y f_\epsilon(x,y) g(y) = g(x), \tag{8.11}$$

if g is a smooth function on Σ. With this we will define a smeared triad as,

$$\left[\tilde{E}_i^3\right]_f (x) = \int_\Sigma d^2 y f_\epsilon(x,y) \tilde{E}_i^3(y), \tag{8.12}$$

which in the limit $\epsilon \to 0$ just goes to $\tilde{E}_i^3(x)$. In terms of these smeared operators the area operator would be given by

$$\hat{A}_\Sigma = \int_\Sigma dx^1 dx^2 \sqrt{\left[\hat{\tilde{E}}_i^3\right]_f \left[\hat{\tilde{E}}_i^3\right]_f}. \tag{8.13}$$

Let us consider the quantization of the smeared triad and its action on a spin network state ψ_s,

$$\left[\hat{\tilde{E}}_i^3\right]_f (x)\psi_s = -8i\pi G\beta \int_\Sigma d^2 y f_\epsilon(x,y) \frac{\delta \psi_s}{\delta A_3^i(y)}. \tag{8.14}$$

Recall that the state ψ_s is a collection of parallel transport operators along open curves tied up by intertwiners at intersections. The key observation here is that the state depends on the connection A_3^i only along the points of the curves. Therefore the functional derivative will only be non-zero if the point y ends up lying on top of the line. That requires that the relevant line of the spin network pierce the area Σ. Otherwise the result is zero. Notice that only that parallel transport operator that happens to pierce the surface will be affected by the derivative (it can happen that more than one pierces the surface, then one will have additional contributions). So the functional derivative can be written as,

$$\frac{\delta \psi_s}{\delta A_3^i(y)} = \mathrm{Tr}\left(\frac{\delta h^J}{\delta A_3^i(y)} \frac{\delta \psi_s}{\delta h^J}\right) \tag{8.15}$$

where h^J is the parallel transport operator associated with the relevant line that pierces the surface in the appropriate J representation. We are using the word "trace" in a generalized sense here: when one differentiates

ψ_s with respect to h^J one pulls h^J out of it and is left with open intertwiners into which the structure of $\delta h^J/\delta A_3^i(y)$ fits in, providing at the end of the day an object with no free indices. We now need to study how to compute the derivative of h^j with respect to the connection. For that we need to go back to the definition of path ordered exponential, equations. (5.23, 5.25),

$$\mathbf{h} = \sum_{n=0}^{\infty} \int_{t_1 \geq \cdots \geq t_n \geq 0} \dot{\gamma}^{a_1}(t_1) \mathbf{A}_{a_1}(t_1) \cdots \dot{\gamma}^{a_n}(t_n) \mathbf{A}_{a_n}(t_n) dt_1 \cdots dt_n. \quad (8.16)$$

Consider the action of the functional derivative on one of the terms of the sum. It will only be non-zero on the integral that touches the point y, assuming such a point lies on the curve γ. Recalling that $\mathbf{A_3} = A_3^i T^i$, where T^i is a basis of the $su(2)$ algebra in the appropriate J representation, we have,

$$\frac{\delta \mathbf{A}_{a_k}(t_k)}{\delta A_3^i(y)} = \frac{\delta \mathbf{A}_{a_k}(\gamma(t_k))}{\delta A_3^i(y)} = T^i \delta^3(y - \gamma(t_k)) \, a\delta_{a_k}^3 \quad (8.17)$$

Therefore,

$$\frac{\delta \mathbf{h}}{\delta A_3^i(y)} = \sum_{n=0}^{\infty} \int_{t_1 \geq \cdots t_k \cdots \geq t_n \geq 0} dt_1 \cdots dt_n \dot{\gamma}^{a_1}(t_1) \mathbf{A}_{a_1}(t_1) \cdots \dot{\gamma}^3$$
$$(t_k) \delta^3(y - \gamma(t_k)) T^i \cdots \dot{\gamma}^{a_n}(t_n) \mathbf{A}_{a_n}(t_n). \quad (8.18)$$

This expression can be broken into three pieces. The first piece, involving the integrals along t_1 to t_{k-1} are just the parallel propagator from the origin of the curve up to the point t_k where the derivative is acting. This is then multiplied times a generator T^i and the tangent to the line at that point. Then it is multiplied times the parallel propagator from the point y to the end of the line. Explicitly,

$$\frac{\delta \mathbf{h}}{\delta A_3^i(y)} = \int dt_k h^J(0, t_k) \dot{\gamma}^3(t_k) T^i \delta^3(y - \gamma(t_k)) h^J(t_k, 1) \quad (8.19)$$

Another way of seeing this is to consider the definition of the exponential as a product, $e^x = \lim_{N \to \infty}(1 + x/N)^N$. A similar definition holds for the path ordered exponential, except that the various factors have to be ordered around the loop. It is clear that the action of the functional derivative will hit generically one of the factors in the middle and the resulting expression will be given by the parallel transport operator up to the point of action from the origin of the loop and will be followed by the parallel transport operator from the point of action to the end of the loop.

Now, one may worry that this action is ill defined. After all, it includes a one-dimensional integral (along t_k) of a three-dimensional Dirac delta. However, we have to remind ourselves that through (8.15) and (8.14) this action is within another two-dimensional integral. One therefore is left with a three-dimensional integral of a three-dimensional Dirac delta function, which is a well defined finite object. We can write the action of the smeared triad as,

$$\left[\hat{E}_i^3\right]_f (x)\psi_s = 8i\pi G\beta\dot{\gamma}^3(x)\mathrm{Tr}\left(T^i\psi_s(x)\right)f_\epsilon(x,y) \tag{8.20}$$

where $\dot{\gamma}^3(x)$ is the tangent of the loop at the point x and ψ_s is the spin network state "opened up" at the point x, so one can insert in that place the generator T^i and retrace. The point y is the place where the surface Σ of the smearing (8.14) intersects s (if it does so in more than one point there is a contribution like this per point and one adds them up). When one acts with the second smeared triad one gets a similar contribution with the insertion of a second generator T^i. Now, noticing that in the $J-th$ representation (meaning with $N=2J$) the product of two generators summed is given by $T^iT^i = J(J+1)1$ (for example, for the fundamental representation where $J=1/2$ one has the Pauli matrices and $S^iS^i = \frac{3}{4}\times 1$ where $S^i = \sigma^i/2$) one gets that the action of the area operator on a spin network state is given by,

$$\hat{A}_\Sigma\psi_s = 8\pi\ell_{\mathrm{Planck}}^2\beta\sum_I \sqrt{j_I(j_I+1)}\psi_s \tag{8.21}$$

where the sum is over all the lines I in the spin network that pierce the surface Σ as shown in Fig. 8.3, and j_I is the "color" of the line I. The presence of G gives rise to the quantity $\ell_{\mathrm{Planck}} = \sqrt{G\hbar/c^3} = 10^{-33}cm$, called the *Planck length* . So we see that spin network states are eigenstates of the area operator and the eigenvalues take a very simple expression reminiscent of the eigenvalues of the operator of the angular momentum squared $\hat{L}^2$ in ordinary quantum mechanics. Notice that the area is quantized in units of the Planck length squared, so the quantum of area is very small even by particle physics standards. The eigenvalues of the area operator are not equidistant, particularly for low values of j_I.

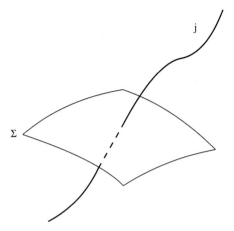

Fig. 8.3 *The surface* Σ *acquires a "quantum of area" when pierced by one of the lines of a spin network.*

Notice that the Barbero–Immirzi parameter appears explicitly in the expression of the area. This means that if we were able to measure the quanta of area we could determine the value of the parameter. In the quantum theory different values of the parameter therefore do not correspond to the same physics. How could this be? Did we not argue that in the classical theory a choice of the parameter is just a choice of canonical coordinates which does not change the physics? This is something that can happen during quantization, that things that are equivalent classically are not quantum mechanically. Usually changes in canonical coordinates (canonical transformations) get implemented in the quantum theory as unitary transformations and therefore do not change the physics. However, that is not always the case and this is what is happening here. Currently we do not have any experiment that could determine β, although we will see in Chapter 10 that when one computes the entropy of black holes, for the results to match semiclassically a particular value of β must be chosen.

The above construction of the area operator highlights key aspects of loop quantum gravity. Notice how the operator is well defined and no divergences appear. Notice the key role of background independence: no background structures are left at the end of the definition. The operator depends on whether the loop intersects the surface and the valences of the line in the loop. It is as if every line in a spin network carries "a quantum of area" and that surfaces acquire their area through piercing of the spin network lines.

In the above derivation we have oversimplified many aspects. In particular, the surface Σ could pierce not only a line in the spin network, but could also contain a portion of a line within itself. It could also contain an intersection of two lines. These cases can be handled with some delicacy and add contributions to the simple formula we derived above. See Ashtekar and Lewandowski (1997) or Thiemann's book (2008) for the complete story.

Another geometric operator of interest is the volume operator. Given a three-dimensional region R, the volume of the region is given by,

$$V(R) = \int_R d^3x \sqrt{\det q} = \frac{1}{6} \int_R d^3x \sqrt{|\epsilon_{gbc}\epsilon^{ijk}\tilde{E}_i^a \tilde{E}_j^b \tilde{E}_k^c|}. \tag{8.22}$$

Promoting this quantity to a quantum operator proceeds along roughly the same lines as in the case of the area operator. One needs to replace the triads by smeared operators. Since the action of each smeared triad is proportional to a tangent to a loop, which in turn are contracted into a Levi–Civita symbol, the only chance of getting a non-vanishing contribution is at a vertex where at least three non-coplanar lines of the spin network intersect. A detailed analysis shows that the action of spin network states can be regularized in a background independent way yielding a finite operator. It is given in terms of a self-adjoint matrix and therefore it is in principle diagonalizable, although a diagonal expression cannot be found in closed form. Moreover, it turns out that one needs at least four-valent vertices for the volume operator to be non-vanishing. The explicit expression is rather complicated and depends on the type of intersections that the spin network has and the valences entering the intersection, so we will not discuss it here. The emergent picture is that spin networks carry "quanta of area" in their links and "quanta of volume" in their intersections and these become the building blocks of a quantum geometry. A given region is endowed with volume if it enclosed vertices of a spin network, and a surface is endowed with area if it is pierced by links of a spin network and one can build a picture of "elementary building blocks" of space as shown in Fig. 8.4. It is clear that in regions where the number of lines and vertices are sparse, one could have weird quantum behaviors (like areas without volumes and volumes without areas). One expects the spacing of the lines in classical geometries to be of the order

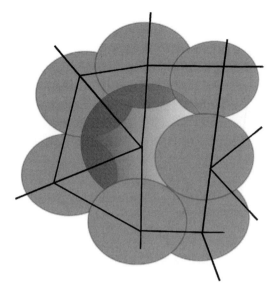

Fig. 8.4 *The spin network can be associated with "elementary building blocks of space-time" in which every vertex of valence four or higher contributes a "volume block" surrounded by areas determined by the valences of the spin network.*

of the Planck length, so one would not see that type of quantum behavior macroscopically.

8.3   The Hamiltonian constraint

We would now like to promote the Hamiltonian constraint to a quantum operator in the loop representation, much like we did with the area operator and argued could be done with the volume operator. The classical expression of the constraint is given by,

$$H(M) = \int d^3x M \left\{ A_c^k, V \right\} F_{ab}^k \epsilon^{abc}. \qquad (8.23)$$

To promote this expression to a quantum operator we will introduce a lattice regularization procedure. Let us assume space has been divided into tetrahedra Δ. We will build an expression such that the limit in which the tetrahedra shrink in size approximates the expression for the Hamiltonian constraint.

For each tetrahedron we pick a vertex and call it $v(\Delta)$. Let $s_i(\Delta)$ with $i = 1, 2, 3$ be the three edges ending at $v(\Delta)$. Now construct a loop $\alpha_{ij}(\Delta) = s_i(\Delta) \circ s_{ij}(\Delta) \circ s_j(\Delta)^{-1}$ by advancing along $s_i(\Delta)$ then along

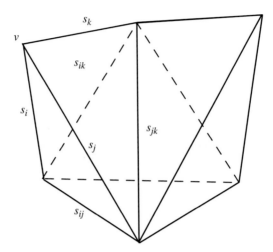

Fig. 8.5 *The "triangulation" used in the regularization of the Hamiltonian constraint consists dividing space into adjoining tetrahedra.*

the line joining the points of s_i and s_j that are not $v(\Delta)$ (we call that line s_{ij}) and then returning to $v(\Delta)$ via s_j, as shown in Fig. 8.5. Let us now consider the parallel propagator along some of the lines of the tetrahedron. For instance, the parallel propagator along the direction s_k. The tetrahedra shrink in size until they become a point (we denote that by the tetrahedron Δ going to the point $v(\Delta)$) one has,

$$\lim_{\Delta \to v(\Delta)} h_{s_k} = \mathbb{1} + \mathbf{A}_c s_k^c \tag{8.24}$$

where s_k^c is a vector in the direction of the edge s_k. The length of such a vector goes to zero in the limit. In the path ordered exponential of the parallel propagator we are therefore left with only the first and second terms in the expansion of the exponential, the other terms being higher order in the length of the edge.

A little more reflection shows that,

$$\lim_{\Delta \to v(\Delta)} h_{\alpha_{ij}} = \mathbb{1} + \frac{1}{2} \mathbf{F}_{ab} s_i^a s_j^b, \tag{8.25}$$

and we see the curvature emerging largely because we are considering a closed circuit, so we are recovering in the limit of an infinitesimal loop the Stokes' theorem that one has for Abelian connections.

Let us now consider the following expression defined on the triangulation,

$$H_\Delta(N) = \sum_\Delta N(v(\Delta)) \epsilon^{ijk} \mathrm{Tr} \left(h_{\alpha_{ij}} h_{s_k} \left\{ h_{s_k}^{-1}, V \right\} \right), \qquad (8.26)$$

where the sum is over all tetrahedra Δ, and see how it behaves when $\Delta \to v(\Delta)$. Starting from the Poisson bracket, we see that the parallel transport within it goes to the identity plus a term proportional to the connection. The identity will have a vanishing Poisson bracket with the volume, so the only contribution will come from the connection. Next we have the parallel transport along s_k outside the Poisson bracket. That would have two contributions, one given by the identity and the other by $A_c s_k^c$. The latter would yield a higher order term since the contribution inside the Poisson bracket is already proportional to s_k^c and we are assuming those to be small. So only the identity contributes. Finally we have that the parallel transport around α_{ij} will yield F_{ab} and an identity term. The identity term vanishes since multiplied times the Poisson bracket it is proportional to a Pauli matrix and that has vanishing trace. We therefore see that the above expression indeed goes in the limit in which the tetrahedra are made infinitesimally small to the Hamiltonian constraint. Notice by the way that all references to a background structure, in particular the length of the links and so on is gone at the end of the process, the three lengths of the s's that appear combine with the summation in the limit to produce an integral.

The reason we went to all that trouble to rewrite the Hamiltonian constraint in terms of parallel transport operators, is that the resulting expression is immediate to promote to an operator in the loop representation. The holonomies are readily promoted to operators, so is the volume operator. A caveat is that one should choose the classical triangulation to be "adapted" to the spin network state one has to act on, by choosing the vertices and lines of the latter to lie on vertices and lines of the triangulation. Note that there will be many lines and vertices in the triangulation that do not correspond to vertices or lines of the spin network when one takes the limit. Due to the presence of the volume the constraint will only receive contributions at vertices of the spin network state considered when there are at least three non-coplanar lines entering the vertex. At such a vertex one can align the i, j, k directions with three directions from the entering lines. The result of the action of the Hamiltonian can therefore be computed. The net effect is to "dress"

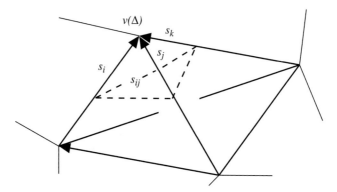

Fig. 8.6 *The triangulation used in the regularization of the Hamiltonian constraint is chosen to be adapted to the spin network state the operator is acting on in such a way that the spin network vertices and edges lie on vertices and edges of the triangulation.*

the vertex with an extra line corresponding to the link s_{ij}, change the valences of s_i and s_j and produce an overall factor, if one is working with trivalent vertices. See Fig 8.6. The action is a bit more complicated for higher valences. Computing the action in detail is elaborate and we will not need it in our treatment, so we refer the reader to the article by Borissov, De Pietri, and Rovelli (1997).

It is worth talking about which space the operator just defined is acting on. The Hamiltonian constraint is not invariant under diffeomorphisms. Therefore its action can only be defined on the "kinematical" space of spin networks not invariant under diffeomorphisms $|k\rangle$. One can transfer its action to diffeomorphism invariant states in the following way. Consider a state $\langle\Psi|$ such that $\langle\Psi|s\rangle = \langle\Psi|s'\rangle$ if the spin networks s and s' are diffeomorphic to each other. Such a state is not really in the kinematical space but is defined as a limit in it, much like ordinary distributions are defined as limits of functions. We then define the action of $\hat{H}(N)$ on Ψ in the following way,

$$\langle\hat{H}(N)\Psi|s\rangle \equiv \lim_{\Delta\to v}\sum_{\Delta\in T}\langle\Psi|\hat{H}_\Delta(N)s\rangle. \qquad (8.27)$$

where the sum is over all the tetrahedra Δ in the triangulation T. This in turn resolves the following issue: if the action of the constraint is to add a line to the spin network state, where precisely is that line added? The answer is that it is irrelevant. When one projects on Ψ the position of the line does not matter. One is working on the space of diffeomorphism invariant states and therefore the line can be slid "closer" or "farther" from the vertex and the action of the operator does not change.

Having a well defined Hamiltonian constraint is a major feat. It constitutes a well defined theory of quantum gravity. Without infinities. This could be considered the major answer of loop quantum gravity to the problems we encountered in Chapter 6 when we discussed perturbative quantum field theory. Here we have a non-perturbative treatment that is finite and devoid of infinities. Thiemann has also shown that if one couples matter to gravity using similar techniques the resulting Hamiltonian constraints are also finite. In a sense, a long held expectation that taking the quantum nature of gravity into account may help regulate the infinities of quantum field is starting to take shape. Finite non-trivial theories of quantum fields are possible within the non-perturbative approach.

Should this be a cause for celebration? Only if the theory can capture the correct physics. It is not meritorious, at least from a physical point of view, to build theories that are mathematically consistent but devoid of physical content. So what can we use to check if the theory has the correct physics in it? Unfortunately this is not an easy question. Presumably one would like to construct physical states, that is, quantum states that are annihilated by the Hamiltonian constraint. But then one is left with the problem of disentangling the true evolution from a theory that is totally constrained. This has not been achieved for general relativity.

What about the algebra of constraints? Can one at least check that the correct algebra is reproduced quantum mechanically by the Hamiltonian constraint that was introduced? Unfortunately this also presents difficulties. In the space of states we are working on, the spin networks endowed with the Ashtekar–Lewandowski measure, the diffeomorphism constraint cannot be represented as an operator. In that space finite diffeomorphisms are well defined but not the infinitesimal ones represented by the constraint. One can represent its solutions as states and therefore one is correctly incorporating the underlying symmetry associated with it, but not the operator itself. Therefore there is no chance of verifying the full constraint algebra since the Poisson bracket of two Hamiltonians is proportional to a diffeomorphism constraint. Moreover, since we defined the action of the Hamiltonian in the space of diffeomorphism invariant states, the commutator of two Hamiltonians should vanish, and indeed it does. Since the action of two successive constraints up to an overall factor is to add two lines, and in the projection on Ψ one is allowed to move lines around, the commutator vanishes. One can also check that the right-hand side of the commutator of two constraints, when promoted to an operator and projected on Ψ also vanishes, and therefore there is no anomaly or inconsistency, at least in this partial sense (Thiemann 2008b).

It should be emphasized that there is quite a bit more technical detail to the action of the constraint, particularly on non-trivalent graphs, and we are being quite sloppy in characterizing precisely which space of states the constraint is acting on. In fact the mere existence of operators like the ones we are discussing requires a lot of mathematical care to establish. We have also omitted all discussion of the difficult looking second term in the Hamiltonian that arises when $\beta \neq i$. As we stated, the same techniques used here can be used to deal with it. We refer the reader to Thiemann (2008) for further discussion, particularly on the types of topologies that need to be considered when computing the commutator of two Hamiltonians and to what extent one can claim the problem is free of anomalies.

Further reading

The books by Rovelli (2007) and Thiemann (2008) cover this material well.

Problems

1. Prove the following Mandelstam identity,

$$W_{\gamma_1}[A]W_{\gamma_2}[A]W_{\gamma_3}[A] = W_{\gamma_1 \circ \gamma_2}[A]W_{\gamma_3}[A] + W_{\gamma_2 \circ \gamma_3}[A]W_{\gamma_1}[A]$$
$$+ W_{\gamma_3 \circ \gamma_1}[A]W_{\gamma_2}[A] - W_{\gamma_1 \circ \gamma_2 \circ \gamma_3}[A] - W_{\gamma_1 \circ \gamma_3 \circ \gamma_2}[A]. \qquad (8.28)$$

2. When discussing the Mandelstam identities, it might have occurred to you to wonder about what happens when the loops γ and η do not have a point in common. How does one define $\gamma \circ \eta$ in that case? Show that the holonomy along a loop is unchanged if one adds to such a loop a retraced path (one parallel propagates along the path away from the loop and then parallel propagates back, as shown in the figure).

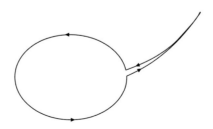

Such retraced paths are called *trees*. By adding such paths one can make any two loops have a point in common.

3. When we promoted the Hamiltonian constraint to an operator in the loop representation, we only considered the first term in the complete expression (7.12). Given the integral $K = \int d^3x K_a^i \tilde{E}_i^a$ and taking into account that $K_a^i = \{A_a^i, K\}$ and that $K = \{H(N = 1), V\}$ where $H(N = 1)$ is the Hamiltonian constraint smeared with a unit lapse, sketch how one could promote the second term to an operator.

4. The action of the Hamiltonian constraint, as discussed in the text, adds a line in the spin network. The spin network therefore has two new vertices, and such vertices are coplanar. Suppose $|s\rangle$ is a spin network with no coplanar vertices. Argue why $\langle s|\hat{H}|s\rangle = 0$.

5. Prove that the formula for Gauss' linking number is one for two linked loops and zero for unlinked ones. Hint: endow each loop with a unit electric current and use Ampère's law.

6. Show explicitly equation (8.25) and show that (8.26), in the limit where Δ shrinks to a point, reproduces the Hamiltonian constraint of the continuum theory.

9
An application: loop quantum cosmology

9.1 The classical theory

As we discussed in Chapter 3, general relativity can be applied to the universe as a whole by approximating the space-time geometry with a homogeneous, isotropic metric. One quickly concluded that models of such a type inevitably encounter a singularity as one evolves them backwards in time: the Big Bang singularity. One of the expectations about quantum gravity is that it will somehow help eliminate the singularities. In this chapter we will study how to apply loop quantum gravity to cosmology. In general, one would have to continue the work outlined in the previous chapter, finding quantum states that are annihilated by the Hamiltonian constraint but that in addition are approximately homogeneous and isotropic. One will not expect, at a quantum level, to find states that are exactly homogeneous and isotropic since that would require freezing degrees of freedom, something the uncertainty principle does not allow in quantum mechanics. Unfortunately, states of this sort are states of the full theory and retain all the complexity of the full theory. So the introduction of isotropy and homogeneity in this way will still be a very challenging undertaking. It will eventually have to be done, but one expects that one could make progress in the isotropic and homogeneous case more quickly using other ideas.

There is an alternative one can explore: what if one reduces the theory to homogeneity and isotropy classically and then proceeds to quantize? If one does that, one is left with a mechanical system, it has a finite number of degrees of freedom. This is what is known as the *mini-superspace* approximation. As in all approximations in physics, one will have to check later that the results one obtains are robust by going outside of the approximation. But it provides a starting point for the analysis. In this limited context, with a finite number of degrees of freedom, one cannot expect to literally deploy the techniques of loop quantum gravity. Moreover, there exists the well known Stone–Von Neumann theorem that states that for systems with finitely many degrees of freedom, all quantum representations are equivalent. People studied quantum cosmologies by reducing and quantizing using traditional variables and found that the singularity was not eliminated (see for instance Kiefer (1988)). Due to the Stone–Von Neumann theorem there was little hope that doing something different could change things within this limited context. So for many years this was not an avenue of research that was actively pursued. But then Bojowald (2000) made the important observation that one could introduce techniques that mimic some key behaviors of loop quantum gravity in the context of homogeneous cosmologies. Moreover, the techniques violated one of the continuity assumptions of the Stone– Von Neumann theorem, opening the door for new results. And remarkably they predicted that the singularity was eliminated in loop quantum cosmology. In this chapter we will review some of these results.

To enforce homogeneity and isotropy we recall that the metric took a very simple form,

$$ds^2 = -dt^2 + a(t)^2 \left(dx^2 + dy^2 + dz^2 \right). \tag{9.1}$$

One can obtain this metric from Ashtekar variables that are diagonal,

$$A_a^i = c\, \delta_a^i, \tag{9.2}$$

$$\tilde{E}_i^a = p\, \delta_i^a. \tag{9.3}$$

Where c and p are functions of time related to $a(t)$ and its time derivative, in particular $a^2 = |p|$, and c is proportional to $\dot{a}(t)$ on solutions of the classical theory as we will see below. One does not have to choose the Ashtekar variables to be diagonal matrices, but we do so here for simplicity (in particular they automatically solve the Gauss constraint). The c and p are canonically conjugate, inheriting the Poisson bracket of A_a^i and $\tilde{E}_i^a$,

$$\{c, p\} = \frac{8}{3}\pi G\beta \tag{9.4}$$

with β the Barbero–Immirzi parameter and G Newton's constant. Notice that we are choosing a different normalization of the densitized triad than in previous chapters, to stay in line with the convention that has become customary in the context of loop quantum cosmology. Given A_a^i one can compute the curvature,

$$F_{ab}^k = c^2 \epsilon_{ab}^k. \tag{9.5}$$

With these we can proceed to write the Hamiltonian constraint,

$$H_G = \frac{\epsilon^{abc} \epsilon_{ijk} \tilde{E}_i^a \tilde{E}_j^b F_{ab}^k}{\sqrt{\det(q)}} + \frac{(\beta^2 + 1)}{\beta^2} \frac{\left(\tilde{E}_i^a \tilde{E}_j^b - \tilde{E}_j^a \tilde{E}_i^b \right) \left(A_a^i - \Gamma_a^i \right) \left(A_b^j - \Gamma_b^j \right)}{\sqrt{\det(q)}} \tag{9.6}$$

$$= -\frac{6}{\beta^2} c^2 \sqrt{|p|}. \tag{9.7}$$

As we mentioned before, this theory in vacuum has flat space as its only solution. To work with something a bit more realistic one has to couple this theory to matter. The matter we will use is the one we discussed in some detail in Chapter 7, a scalar field. Except that here the scalar field we will consider is homogeneous. The diffeomorphism constraint vanishes automatically due to homogeneity. The contribution to the Hamiltonian constraint for a field without a potential is very simple $H_\varphi = 8\pi G p_\varphi^2 / |p|^{3/2}$. In this context we are dropping all tildes on densities since we think of the variables as only functions of time as in a mechanical system. One has a Hamiltonian constraint $H = (H_G + H_\varphi)/(16\pi G)$, where we added an overall denominator to better match the literature, it is equivalent to rescaling the lapse.

Let us work out the classical dynamics of this cosmology. Since φ does not appear in the Hamiltonian, p_φ is a constant of the motion. Choosing from now on a unit lapse $N = 1$, the equation of motion for φ is $\dot{\varphi} = p_\varphi / p^{3/2}$. The equation of motion for p is,

$$\dot{p} = \{p, H\} = -\frac{8\pi \beta G}{3} \frac{\partial H}{\partial c} = \frac{2\sqrt{|p|}c}{\beta}. \tag{9.8}$$

Since $H = 0$ one has that,

$$c = \frac{2\sqrt{3\pi} \beta G}{3} \frac{p_\varphi}{p}, \tag{9.9}$$

and therefore we can write

$$\dot{p} = \frac{4\sqrt{3\pi}G}{3}\frac{p_\varphi}{\sqrt{|p|}}, \tag{9.10}$$

which can be immediately integrated to give,

$$p^{3/2} = \frac{8\sqrt{3\pi}G}{3}p_\varphi t, \tag{9.11}$$

which shows that the volume of the universe goes to zero at $t = 0$, giving the Big Bang singularity. Conventional texts in cosmology rewrite the above equation of motion in terms of the *Hubble parameter*[1], $\mathcal{H} = \dot{p}/(2p)$, and the matter density of the scalar field $\rho = p_\phi^2/(2|p|^3)$ as,

$$\mathcal{H}^2 = \frac{8\pi G}{3}\rho, \tag{9.12}$$

known in the literature as one of *Friedmann's equations*. We will see that loop quantum cosmology implies a modification to this equation.

9.2 Traditional Wheeler–De Witt quantization

Up to now we have only made a change of variables in a system with a finite number of degrees of freedom so obviously nothing changes. Interesting things will start to happen when we quantize the system using ideas parallel to those of loop quantum gravity. If we were to just consider the above mechanical system with phase space variables $(c, p, \varphi, p_\varphi)$ and just proceed to quantize it, we would recover traditional quantum cosmology. Let us explore briefly how this works. We choose a p, φ representation and we have therefore that $\hat{p}$ and $\hat{\varphi}$ act multiplicatively on states according to,

$$\hat{c}\Psi = i\hbar\frac{8\pi\gamma G}{3}\frac{\partial\Psi}{\partial p} \quad \text{and} \quad \hat{p}_\varphi\Psi = -i\hbar\frac{\partial\Psi}{\partial\varphi}. \tag{9.13}$$

The quantum version of the Hamiltonian constraint can be readily written, it is known as the *Wheeler–De Witt equation*. The form of the Hamiltonian constraint immediately implies that p_φ is a constant of the

---

[1]Sometimes referred to as *Hubble constant* though it is really a function of time.

motion, therefore the field φ is a monotonic function on each classical trajectory. Therefore one can consider it as a "time" and the Wheeler–De Witt equation can be viewed as an evolution equation on this *emergent time* φ. In this context p_φ is a Dirac observable. There is another Dirac observable which corresponds to the volume of the universe at a given time V_φ. This is a complete set of Dirac observables for the model. Requiring that the Dirac observables be self adjoint uniquely fixes the inner product. We will not work out the details here but refer the reader to the paper by Ashtekar, Pawlowski, and Singh (2006). One can then construct wave-packets that are peaked around a classical solution, say, for large volumes of the universe with small uncertainties in the volume and its canonically conjugate variable and study the evolution of the state as a function of our clock φ. The end result is that the uncertainty remains small all the way to the Big Bang and therefore the expectation that large quantum effects near the singularity will somehow change things does not materialize and the singularity is not removed by the quantization.

9.3 Loop quantum cosmology

We will now recall two key elements of loop quantization: the basic elementary quantum operators are given by the holonomy and the triad flux operators. Since space is homogeneous we can consider an "elementary cell" that is repeated over and over again and just concentrate on studying the dynamics of that cell. Let us construct the holonomy along a loop going around the cell of the connection $A_a^i = c\delta_a^i$. Notice that this is really an Abelian holonomy since it is proportional to the identity matrix and therefore one can just consider the ordinary integral of c along the line. Notice that due to homogeneity, that integral is just c times the length of the edge of the elementary cell, we will call it λ. Therefore the holonomy would be given by $h_\lambda = \exp(i\lambda c)$. Some papers in the literature keep the volume of the elementary cell explicit since it is a free parameter. Here we will assume the volume has been chosen so the expressions have their simplest form. In the loop representation the volume will be promoted to a well defined operator. On the other hand the variable c, which plays the role of a connection, does not exist as a quantum operator in the loop representation. Although the quantity p, which plays the role of the triad, does exist as a quantum operator (due to homogeneity it is not terribly relevant to make a distinction between the triad and the triad flux operator). This may appear surprising. Could we not define an operator $\hat{c}$ by considering the derivative of $\hat{h}_\lambda$ with respect to λ? The answer is yes, if the representation of $\hat{h}_\lambda$ were continuous with respect to λ. However, the one we will choose, in parallel to what happens in full loop quantum gravity, will not be continuous. That, incidentally, is the hypothesis of the Stone–Von Neumann theorem that is violated here; this allows something non-trivial to happen. It introduces in this simple quantum mechanical context new representations that capture the flavor of what is going on in loop quantum gravity.

This may appear far fetched: why have we never encountered such representations in quantum mechanics before? In ordinary quantum mechanics these representations are indeed unnatural since we need the operator $\hat{x}$ directly, not its exponential. In loop quantum gravity, as we argued, it is essentially the *unique* representation available. The two contexts are dramatically different, largely due to diffeomorphism invariance, and therefore it is not surprising that different representations should arise. What about cosmology? Since cosmologies are mechanical systems one has a choice: one could treat them the usual way as we did in the previous section. But such a treatment is unlikely to parallel behaviors from full loop quantum gravity. For cosmological models, in spite of their being mechanical systems, it is more natural to consider the types of representations we will use in this section.

Let us proceed to build the representation. We introduce a kinematical Hilbert space, just like in ordinary quantum mechanics, so we can speak concretely about operators. On such a space $\hat{p}$ defined as a multiplicative operator is naturally self-adjoint. We can describe the gravitational part of kinematical Hilbert space in terms of the eigenbasis of $\hat{p}$. A general state $|\Psi\rangle$ has the form,

$$|\Psi\rangle = \sum_i \Psi_i |p_i\rangle \tag{9.14}$$

and $|p_i\rangle$ is an orthonormal basis,

$$\langle p_i | p_j \rangle = \delta_{ij}. \tag{9.15}$$

Up to this point everything looks like familiar quantum mechanics, say, applied to the free particle or the harmonic oscillator. But there is a big difference: the state is a *countable* sum of eigenstates rather than an integral and in the inner product one has a Kronecker delta rather than a Dirac delta. So this space is not endowed with the usual inner product on square integrable functions. This of course mimics better the situation of the inner product of loop states given by the Ashtekar–Lewandowski measure. But, as we will immediately see, it has profound consequences. The action of the basic operators is as follows,

$$\hat{p}|p\rangle = p|p\rangle \quad \text{and} \quad \hat{h}_\lambda|p\rangle = |p + \lambda\rangle \tag{9.16}$$

just like $\hat{p}$ and $\exp(i\lambda x)$ in ordinary quantum mechanics. However, is not *continuous* in λ since $|p + \lambda\rangle$ is orthogonal to $|p + \lambda'\rangle$ for all $\lambda \neq \lambda'$. Consequently one cannot take the derivative of $\hat{h}_\lambda$ with respect to λ, which would allow one to define the operator $\hat{c}$. As we stated, this lack of continuity is the hypothesis of the Stone–Von Neumann theorem that is violated by quantizations of the style of loop quantum gravity. Even in such a simple setting as cosmology.

9.4 The Hamiltonian constraint

Now that we have the kinematics under control we need to construct a well defined operator representing the Hamiltonian constraint. We cannot simply promote the expression (9.7) to a quantum operator since there is no operator associated with c. So we consider another expression[2],

$$16\pi G H_{\text{effective}} = -\frac{6}{\beta^2}|p|^{\frac{1}{2}}\frac{\sin^2(\mu_0 c)}{\mu_0^2} + 8\pi G \frac{1}{|p|^{\frac{3}{2}}}p_\varphi^2 \tag{9.17}$$

where we have made the replacement $c \to \sin(\mu_0 c)/\mu_0$. The above expression would therefore coincide with the Hamiltonian constraint in the limit $\mu_0 \to 0$. The above expression can be readily represented as a quantum operator, since the sine function is easily expanded in terms of the holonomies h_{μ_0}. The tricky part, however, is that the resulting quantum expression will not allow us to take the limit $\mu_0 \to 0$ due to the natural discreteness of the loop representation. The quantum theory, unsurprisingly, will fail to have ordinary general relativity, as a semiclassical limit at least in certain regimes. So if one cannot study the limit μ_0 what does one do with such a parameter? It was introduced artificially, how can one

---

[2]Alternatively, and more correctly, one can obtain the quantum Hamiltonian constraint following a procedure similar to that introduced by Thiemann in the full theory. The final result is the same, see Ashtekar, Pawlowski, and Singh (2006).

assign a value to it? The idea here is get to the root of why one cannot take the limit in the quantum theory. One is replacing the quantity c with a holonomy h_{μ_0}. The limit $\mu_0 \to 0$ is equivalent to shrinking the loop around which the holonomy was computed down to a point. In this very simple context of loop quantum cosmology the only thing left is the length of the loop μ_0 going to zero. In the full theory one would have a real loop with a real area encompassed by it. Since we know that areas are quantized in loop quantum gravity it is not surprising that one cannot shrink the loop to zero. That would suggest as a natural value for μ_0 the smallest eigenvalue of the area operator, $\ell^2_{\text{Planck}}\beta$. This corresponds to $\mu_0 = 3\sqrt{3}/2$.

9.5 Semiclassical theory

We will not study the full quantum theory here, we refer the reader to Ashtekar, Pawlowski, and Singh (2006) for the full treatment. What we will do here is a procedure that has proved quite useful. We will study the classical dynamics of the effective Hamiltonian (9.17) using the value of μ_0 we discussed in the previous paragraph. It is clear that such a Hamiltonian is what will be captured in the semiclassical limit of the quantum theory. So we work out the equations of motion,

$$\dot{p} = \{p, H_{\text{effective}}\} = -\frac{8\pi\beta G}{3}\frac{\partial H}{\partial c} = \frac{2|p|^{\frac{1}{2}}}{\beta\mu_0}\sin(\mu_0 c)\cos(\mu_0 c). \qquad (9.18)$$

These can be rearranged into a modification of the Friedmann equation (9.12),

$$\mathcal{H}^2 = \frac{\dot{p}^2}{4p^2} = \frac{8\pi G}{3}\rho\left(1 - \frac{\rho}{\rho_{\text{crit}}}\right), \qquad (9.19)$$

where the critical density is

$$\rho_{\text{crit}} = \left(\frac{3}{8\pi G\beta^2\mu_0^2 p}\right). \qquad (9.20)$$

As we see, the expression for $\mathcal{H}^2$ is not always positive definite. In particular it is zero when,

$$\rho = \rho_{\text{crit}} = \left(\frac{3}{8\pi G\beta^2\mu_0^2}\right)^{\frac{3}{2}}\frac{\sqrt{2}}{p_\varphi}. \qquad (9.21)$$

So if one runs the evolution of the universe backwards, eventually one reaches a point where $\mathcal{H}^2$ vanishes. From there on if one continues backwards the universe re-expands. A "bounce" has replaced the Big Bang. Notice that the value of ρ_{crit} depends on p_φ, which is arbitrary. That means that the bounce could occur at an arbitrary density. This was a serious drawback of initial attempts to build a loop quantum cosmology. It was fixed with a refinement of the construction in which the μ_0 parameter is taken to depend on the canonical variables (Ashtekar, Pawlowski, and Singh 2007). The form of the equation remains unchanged but now the critical density is fixed to $\rho_{\text{crit}} \sim \rho_{\text{Planck}} \sim 10^{100} kg/m^3$ is where the "bounce" occurs, that is the contraction ceases. The value corresponds to a fraction of the Planck density. So we see that for densities much smaller than the Planck density, the theory approximates general relativity well. However, as we follow the evolution back into the past and the density increases, eventually there are significant departures from general relativity. Notice that the term ρ/ρ_{crit} contributes with a different sign to the speed of contraction of the universe. Therefore it is unsurprising that it is actually conspiring to eliminate the Big Bang. What is the origin of this repulsion? One can picture it as a theory with a minimum quantum of area and volume refusing to shrink beyond the smallest quantum. Of course, the analysis we are carrying out is a bit misleading. We claimed that we were studying a semiclassical approximation to the theory. That approximation is very likely not going to be good at regimes where $\rho \sim \rho_{\text{crit}}$. So one cannot seriously conclude from this analysis that the Big Bang is eliminated. However, that is indeed the conclusion of the full quantum treatment of loop quantum cosmology. Such treatment can be seen in Ashtekar, Pawlowski, and Singh (2007): the Big Bang is replaced by a Big Bounce and the cosmology starts expanding again into the past. Also, as we mentioned, in modern treatments like this one it is concluded that μ_0 should not be a constant as we argued here but a dynamical variable, since the length of the elementary loop is obviously metric-dependent and the metric is a dynamical variable. Taking this into account actually improves the behavior of loop quantum cosmologies even further. For instance ρ_{crit} becomes independent of the canonical variables, it just becomes a combination of fundamental constants. That means that the density at which quantum gravity effect becomes important is fixed for all possible solutions, which is more satisfactory than having it depend on the solution chosen.

We hope to have given here a flavor of what is going on without having to go into all the details of the full quantum treatment and all the latest developments. In particular one should be cautious. Loop quantum cosmology is not automatically an approximation to some regime of the full theory but rather a truncation of the full theory. The only way to ensure that one is approximating some regime of the full theory with the simple models considered is to consider models of more and more complexity and show that the results obtained survive. The models studied up to now are very simple ones and one should not infer that it has been established as a fact that singularities are resolved by loop quantum gravity. There is ongoing work in anisotropic models (for example Bojowald (2004), Ashtekar and Wilson-Ewing (2009)) of cosmology and some partial results on models where the metric has spatial dependence (Garay *et al.* (2010), Gambini and Pullin (2008b)) that seem to confirm the results obtained in the simple homogeneous models, although it is premature to draw conclusions from this. More work, particularly in inhomogeneous models, will be needed to verify that the hints gathered from these very simple models indeed bear relation to behaviors of the full theory.

Further reading

The review articles by Ashtekar and Singh (2011) and Singh (2011) provide a pedagogical introduction to loop quantum cosmology. Good coverage is also in the *Living Reviews* article by Bojowald (2008).

Problems

1. Derive the expression for the Hamiltonian constraint in the Friedmann–Robertson–Walker space-time.
2. Using Ashtekar, Pawlowski, and Singh (2006) as guidance, show that wave-packets that can be constructed in the Wheeler–DeWitt theory have a singularity.
3. Study numerically the effective equations stemming from loop quantum cosmology for the model discussed in this chapter and show that the singularity is avoided.

10

Further developments

In spite of the difficulties of dealing with the Hamiltonian constraint and of defining observables that represent physical quantities, some degree of progress has been possible in certain special circumstances. Here we summarize briefly some of these results. A detailed discussion of most of them exceeds the scope of this book, but we present a picture for completeness of the state of the art of the field.

10.1 Black hole entropy

10.1.1 Black hole thermodynamics

In Chapter 3 we introduced the idea of a black hole, a region of space-time that if one enters it, one cannot escape from. It was related with a solution of the Einstein equations, the Schwarzschild solution.

$$ds^2 = - \left(1 - \frac{2GM}{c^2 r}\right) dt^2 + \left(1 - \frac{2GM}{c^2 r}\right)^{-1} dr^2 + r^2 \left(d\theta^2 + \sin^2 \theta d\varphi^2\right).$$
$$(10.1)$$

The region $r < 2M$ is the region that is trapped and the surface $r = 2M$ is called the *event horizon*. It took only one year for this solution to be generalized to a black hole with electric charge, yielding the Reissner–Nordström solution. It took almost 50 years more to find a solution that represents a black hole that has angular momentum (the Kerr solution) and more time to find a solution with angular momentum and electric charge (the Kerr–Newman solution). Remarkably, it was shown (Israel 1967) that the spherically symmetric black holes were uniquely given by the Reissner–Nordström solution. This result was later extended (Carter 1971) to the axisymmetric case showing that the Kerr–Newman solution was the unique solution. That is, any black hole can be characterized by three parameters, its mass, electric charge, and angular momentum. This is a remarkable result. Compare with the situation in

Newtonian gravity. There all masses have a gravitational potential that, to leading order, goes as $1/r$, but in higher powers of the $1/r$ expansion one can have infinitely many different parameters that correlate with the shape of the body in question. You may ask: how can this be? If one collapses a deformed star (say one that is the product of a collision of two stars) into a black hole, does it yield the same black hole as a spherical star? Or suppose you form a black hole and later on drop a star into it. Is the final black hole identical to the one that started the process (perhaps with more mass) but no other distinguishing features? The answer is yes. When something non-spherical collapses into a black hole, initially the hole is highly distorted, but the horizon starts to vibrate, emitting gravitational waves in the process. Eventually all additional information is radiated away and one is left with a Schwarzschild black hole asymptotically into the future if there was no angular momentum or electric charge in the initial configuration.

Given a black hole with mass M, angular momentum L, and charge Q one can drop charged particles into it and alter the mass, charge, and angular momentum. In particular, rather surprisingly, one can extract energy from a black hole through the *Penrose process*. By carefully selecting how one drops particles, one can reduce the black hole angular momentum or its charge and therefore lower its total energy. However, there is a maximum amount of energy one can extract that way. Christodoulou and Ruffini (1971) showed that there exists an *irreducible mass* of a black hole of a given mass, angular momentum, and charge that one cannot reduce beyond. Its expression is given by (in units where $G = c = 1$),

$$M_{\text{irr}} = \frac{1}{2}\sqrt{\left(M + \sqrt{M^2 - Q^2 - a^2}\right)^2 + a^2}. \qquad (10.2)$$

where $a = L/M$ is the Kerr parameter. One cannot extract energy from an uncharged, non-rotating black hole and $M_{\text{irr}} = M$. An interesting observation is that the area of the black hole horizon on a spatial surface is given by $A = 16\pi M_{\text{irr}}^2$. So the area of the horizon of a black hole cannot be diminished by disturbing the black hole. It is condemned to grow. Hawking (1971) actually proved independently and in very general terms that the area of the black hole cannot be decreased by any process. This result implies, among other things, that if two black holes collide and merge, the resulting black hole has an area at least equal to the sum of the areas of the merging holes. This led several authors, but mainly

Bekenstein (1973), to argue that the area of a black hole behaves like the entropy of a thermodynamical system.

At first it may appear far fetched to make an analogy between the non-decreasing nature of a black hole's area and that of entropy. After all, we have argued that a black hole is the analog in general relativity of a mass in Newton's gravity. One would not assign an entropy to such a system. But this is where the differences between the Newtonian and general relativistic cases matter. In the case of black holes we argued that all of them are characterized by three parameters, M, Q, and L. The presence of the event horizon prevents us from gaining any knowledge about all the matter that went into the formation of the black hole apart from those three parameters. Entropy is a notion associated with lack of information about a system. Such a property is clearly present here. Also, it can be said that in usual thermodynamics an increase of entropy is associated with a degradation of the energy of the system, in the sense that some of it cannot be transformed into work. Similarly, in the case of black holes we saw that the irreducible mass cannot be extracted from the black hole by any process, meaning a portion of the energy of the system cannot be turned into work. Therefore an increase in the area (and therefore in M_{irr}) implies the energy of the black hole has degraded. In the case of a Schwarzschild black hole the mass equals the irreducible mass and therefore none of it can be turned into work. It is the analogue of a thermodynamical system in equilibrium. However, one can collide two Schwarzschild black holes and extract work from them in the form of radiated gravitational waves. It is the analogue of two systems individually in equilibrium but that do work when they are brought into thermal contact with each other.

One can push the thermodynamical analogy further, for instance by developing a first law of thermodynamics for a black hole. Recall that in ordinary thermodynamics the first law states that,

$$dE = TdS - PdV \tag{10.3}$$

where dE is the change in energy, dS the change in entropy, dV the change in volume, T the temperature, and P the pressure. To develop the first law for black holes let us rewrite the expression for the area of the horizon a bit using the definition

$$r_{\pm} = M \pm \sqrt{M^2 - Q^2 - a^2}, \tag{10.4}$$

as

$$\frac{A}{4\pi} = r_+^2 + a^2 = 2Mr_+ - Q^2 \tag{10.5}$$

where we have used the definition of the irreducible mass (10.2), its relation to the area, and the last step is just an algebraic identity. In the case of charged and rotating black holes there is an "exterior" horizon r_+ which corresponds to the "surface of no return" but there also is an "interior" horizon r_- that corresponds to a surface one cannot evolve the geometry beyond even if given all the possible initial data in the universe (such horizons, called *Cauchy horizons* are known to be unstable). If we differentiate this expression and solve for dM one gets,

$$dM = \Theta \frac{dA}{4\pi} + \vec{\Omega} \cdot d\vec{L} + \Phi dQ \tag{10.6}$$

where $\Theta = (r_+ - r_-)/(4(r_+^2 + a^2))$, $\vec{\Omega} = \vec{a}/(r_+^2 + a^2)$, and $\Phi = Qr_+/(r_+^2 + a^2)$. This yields in the black hole context an analogue of the first law of thermodynamics. The terms $\vec{\Omega} \cdot \vec{L}$ and ΦdQ are readily interpreted as the PdV work term and would correspond to the work done on a black hole by an infalling particle or other disturbing agent that increases the charge of the black hole by dQ and the angular momentum by $d\vec{L}$. The most remarkable thing is the TdS term is now clearly indicating what we had hinted at before: it identifies the entropy of the black hole as $S = A/(4\pi)$. The only unnatural element at this point is the role of Θ as a temperature. What could it possibly mean to assign a temperature to a classical field configuration as a black hole?

10.1.2 Hawking radiation

In 1975 Hawking made a remarkable discovery that fitted perfectly into the black hole thermodynamical analogy. By studying the propagation of quantum fields on the background of a black hole he noted that a black hole emits thermal radiation at a temperature precisely given by the quantity Θ that appears in the first law. A complete derivation of Hawking's result is too lengthy for the confines of this book and can be seen, for instance, in the book by Carroll (2003). Here we would like to give an idea of where the effect emerges from. To do this we need to briefly consider quantum field theory on a curved space-time. To do this, let us retrace the steps followed in Section (6.2) but for a scalar field theory on a curved space-time. The Lagrangian, as we discussed in Section 7.3 will be

$$L = \int d^3x \sqrt{-\det(g)} \left(-g^{\mu\nu} \partial_\mu \varphi \partial_\nu \varphi - V(\varphi) \right), \qquad (10.7)$$

where we see the main modification is that the metric is not the Minkowski metric anymore and we need to explicitly include its determinant in computing the integral. From this Lagrangian one can compute the equation of motion and one obtains the wave equation on a curved space-time, and we particularize to the case of a free scalar field with mass m, in which $V(\varphi) = m^2 \varphi^2$,

$$g^{\mu\nu} \nabla_\mu \nabla_\nu \varphi - m^2 \varphi = 0. \qquad (10.8)$$

We can proceed to define canonical variables exactly as we did in Section (6.2) with the same canonical commutation relations. We can then proceed to promote the variables to quantum operators. The next step is when things change. That step was to do a Fourier decomposition that allowed us to write the fields in terms of positive and negative frequencies as in (6.33). We could do a Fourier decomposition here, but the modes will get coupled since the equation of motion has changed to (10.8). There is an alternative: the Fourier decomposition is really an expansion of a function in a basis of functions given by $\exp(ik_\mu x^\mu)$. Such functions are not solutions of (10.8). Could one do an expansion in a basis of solutions of that equation? The answer is in the affirmative, one can find a basis of solutions and perform the mode expansion there. Say the basis of functions is given by f_i (the index i may be continuous or discrete depending on the character of the equation (10.8) for a given metric $g^{\mu\nu}$). The fields would then read,

$$\hat{\varphi} = \sum_i \hat{a}_i f_i + \hat{a}_i^\dagger f_i^* \qquad (10.9)$$

and if the basis of functions is orthonormal in a suitable inner product that is preserved by the evolution implied by (10.8), one would still have the usual commutation relations for the creation and annihilation operators and could construct a vacuum just like we did in the flat space-time case. Would that suffice? The answer is yes with an important caveat. The construction is highly non-unique. One could have chosen a different family of orthogonal modes g_i and carried out the construction. The vacuum state in one of the constructions would not correspond to a vacuum state in the other construction. In flat space-time there is a preferred set of coordinates and a preferred choice of time, and by using it, a preferred definition of the vacuum. This is lost in a generic curved space-time.

The problem even extends to the definition of particles in flat space-time when one allows non-inertial observers. As we well know, accelerated observers in flat space-time will most naturally see a metric that is apparently curved (For example the Ehrenfest carousel). Only by computing the curvature can they know they are in flat space-time. That means that if we go through the above construction there will be a non-trivial $g^{\mu\nu}$ present and again the non-uniqueness of the vacuum will arise. That is, an accelerating particle detector will detect particles even if a particle detector in an inertial frame indicates we are in the vacuum. This phenomenon is called the *Unruh effect*.

A similar effect is at the heart of the observation of Hawking that black holes radiate. Consider an observer at rest outside a Schwarzschild black hole at a given radius. Such an observer is not inertial, an inertial observer would free-fall into the black hole. Assuming that the natural place to define a vacuum is in an inertial frame, when we translate back to the frame of an observer at a fixed radius, such an observer will see particles. These are the particles of the Hawking radiation.

A more pictorial way of addressing the Hawking radiation is to consider the virtual creation of a pair of particles like the ones we discussed in Fig. 6.2. Suppose such a pair were produced close to the horizon of the black hole and one of the particles fell in whereas the other escaped to infinity. It would then give the impression that the black hole is radiating. This is really an informal picture, the correct explanation is the one we gave before, but it helps give some intuitive feel for what is happening. It can also be somewhat validated by computing the expectation value of the stress energy tensor of a field on the horizon and noting that there is a negative flux of energy entering it (Frolov and Novikov 1998). This, by the way, would not happen if one did not have a horizon. So for instance, for a very compact neutron star, although the geometry outside the star may resemble somewhat that outside a black hole, there is no Hawking radiation.

An interesting aspect is to consider the Hawking temperature for a Schwarzschild black hole. If one restores all physical constants we set to one and insist on measuring entropy in thermodynamical units, which introduces Boltzmann's constant k_b, one has that

$$T = \frac{\hbar c^3}{8\pi GMk_b} = 10^{-6} \left(\frac{M_{\text{Sun}}}{M}\right) {}^\circ K \qquad (10.10)$$

so for a solar sized black hole the temperature is $10^{-6}\,{}^\circ K$, too small to be observed. Remarkably, the temperature is inversely proportional to the mass. The presence of the outgoing radiation implies that the black hole "evaporates," losing mass, and eventually it will disappear completely. For solar sized black holes that process is too slow to be observed. However, notice that as the black hole loses mass, its temperature increases and the process accelerates.

The introduction of the Hawking radiation in a sense completes the thermodynamic picture of a black hole but also opens new questions. The rest of the thermodynamical picture was entirely classical, yet the temperature arises only when quantum field theory in curved space-time is brought to bear. Would it not be reasonable that some of the other elements in the description should have a quantum origin as well? It is instructive for this to consider the entropy of a black hole of mass equal to the Sun. The entropy of a black hole is proportional to the area. The entropy in the units we are working with is a dimensionless number, so the area has to be divided by a set of fundamental constants that yield an area, which is given by Planck length squared. That means that, for instance, in meters squared $S \sim \text{Area} \times 10^{70}$. Suppose we consider the entropy of a black hole with mass equal to that of the Sun. Such black holes have a Schwarzschild radius of about one kilometer. Their area is therefore of the order of $\times 10^6$ meters squared. As a consequence the entropy of a solar sized black hole is 10^{76} in dimensionless units. This is a huge number. It is about 22 orders of magnitude larger than the thermodynamic entropy of the Sun.

The last observation suggests that the entropy of a black hole cannot be given simply by the entropy of the matter that formed it. Moreover, both the expression of the entropy and of the temperature involve Planck's constant. This suggests that both should have a quantum origin. We already know that the temperature was a quantum effect in nature. But what about the entropy? Could quantum gravity account for the entropy of a black hole? We will discuss this in the next subsection.

Before ending it is worth mentioning another reason why the picture involving quantum field theory in curved space-time is not entirely satisfactory and why eventually a full quantum gravitational description will be needed. As we observed, as the black hole radiates, its mass diminishes, which in turn implies that its temperature increases. As the black hole becomes smaller, the curvature at the horizon gets larger and larger (the horizon gets closer to $r = 0$ and fields behave as inverse powers of the radius). As the curvature becomes larger it becomes less and less tenable to ignore quantum fluctuations of the gravitational field itself. In particular, if one insists in carrying out the semiclassical picture all the way to the end one encounters inconsistencies. The most famous is the *information paradox*. All of the information about the matter that formed the black hole was presumably trapped inside the event horizon. If the black hole evaporates and all that is left is thermal radiation going out (characterized by a single number, its temperature), what happened to all that information? We know in quantum mechanics that evolution of states is given by a unitary evolution operator. This implies that all information present in an initial state can be retrieved in the final state since they are related by a unitary transformation. One might say, what is different here from, say, tossing a book into a fire? Is the information not lost there too? The answer is negative. The radiation from the fire is not thermal, and conceivably by carefully observing the flames one could reconstruct the information lost. This sounds far fetched and it is, but one has to consider that the radiation in the black hole is perfectly thermal, meaning completely characterized by the number T and therefore one cannot reconstruct the information from it. Solving this apparent puzzle is one of the major goals of all quantum gravity programs.

To conclude this section it is worth mentioning that taking the thermodynamical ideas very seriously has led several authors to argue that general relativity follows from thermodynamics. This is an intriguing idea but at the moment loop quantum gravity does not seem to play a role in it so we will not discuss it here. For further details see Jacobson (1995) and Padmanabhan (2010).

10.1.3 Black hole entropy in loop quantum gravity

Starting with suggestions by Krasnov (1997) and Rovelli (1996) and further developed by Ashtekar, Baez, Corichi, and Krasnov (2000) attempts

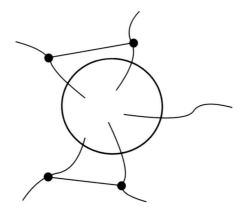

Fig. 10.1 *The black hole horizon acts as a boundary of space-time which is punctured by the lines of the spin network.*

to explain black hole entropy in loop quantum gravity have been studied by several authors. All these calculations operate in the approximation that the black hole is large (that is not Planck sized). This is reasonable since for small black holes there will be a lot of Hawking radiation present and the mass would be evolving rapidly. In fact, the loop quantum gravity calculations up to now ignore Hawking radiation entirely. The idea is to consider a space-time that classically contains a boundary. The boundary is the horizon of the black hole and one studies the exterior of the black hole and the boundary and how they interact, as shown in Fig. 10.1. We will not be able to give a detailed picture of the calculation, but will outline the broad elements that go into the calculation in this section. The novel element is the presence of the boundary, which we call S. To begin with, one has to impose that the boundary correspond to a horizon. A framework developed in recent years called *isolated horizons* (Ashtekar and Krishnan 2004) allows us to do that. Starting with the spin connection Γ_a^i, and defining its pull back onto S as $\bar{\Gamma}_a^i$, and picking a constant vector in the internal indices r_i, one can define an Abelian connection on the boundary S, $W_a = -\bar{\Gamma}_a^i r_i$. If one defines an antisymmetric tensor with internal indices as $\Sigma_{ab}^i \tilde{\epsilon}^{abc} \equiv \tilde{E}^{ci}$, and pulls it back to the surface, which we again denote it by an overbar, the condition of having an isolated horizon is given by

$$\partial_a W_b - \partial_b W_a = -2\beta \bar{\Sigma}_{ab}^i r_i. \tag{10.11}$$

The second issue in dealing with a field theory in a manifold with a boundary is that when one computes the functional derivatives to compute the equations of motion from the action and the canonical momenta, the action fails to be differentiable at the boundary. This requires correction, which in turn yields a separate canonical theory at the boundary in addition to the usual field theory in the "bulk" of the space-time. The theory in the boundary has a constraint given by (10.11).

Another condition that has to be met for the boundary to be an isolated horizon is that the lapse vanishes at the boundary. That means that the Hamiltonian constraint does not generate any evolution there and the horizon is "isolated." If it did not vanish it would mean that the black hole is evolving (that is it is swallowing matter) and therefore it would not be isolated. This is important because it automatically turns the area of the horizon into a Dirac observable. Since as we know the entropy should be proportional to the area, it is reassuring that the proportionality is with a quantity that is physically meaningful in the canonical theory. The vanishing of the lapse also means that we do not need to concern ourselves with the complex problem of the dynamics of loop quantum gravity in the bulk, since the entropy will become dependent only on quantities in the boundary.

To define the entropy let us assume that one is given a black hole of area a_0. We will take the microcanonical viewpoint. We will count the number of quantum states N such that the eigenvalue of the area operator is in the an interval of width 2δ around a_0. We will require that the condition (10.11) be satisfied by the states. The entropy will be given by the logarithm of N. To count the states we note that if a link with valence j_I from a spin network in the bulk touches the horizon it will endow it with an area $A = 8\pi\beta\ell_{\text{Planck}}^2 \sqrt{j_I(j_I+1)}$. At a puncture the operator associated with $\bar{\Sigma}_{ab}$ will also act. Notice that this operator is essentially the component of the triad that is orthogonal to the surface of the horizon, similar to the $\tilde{E}^3$ operator we considered when we constructed the eigenvalues of the area operator. Such an operator we noted played the role of $\vec{J}$, an angular momentum operator, whereas the area itself was related to $\sqrt{\vec{J} \cdot \vec{J}}$. The eigenvalues of $\hat{\Sigma}_{ab}$ are half integers m_I that satisfy $-J_I \leq m_I \leq J_I$. The requirement that the surface of the horizon is topologically a sphere implies that $\sum_I m_I = 0$. This can be roughly understood as "strands that enter must exit." Therefore a quantum state of the horizon is characterized by a set of punctures p_I where the spin network from the bulk penetrates it together with a pair of half integers (J_I, m_I) per puncture. One has yet to implement diffeomorphisms that keep the boundary fixed. If one has two pairs of punctures with the same labels and one performs a diffeomorphism that exchanges them, does that count as a new state? A careful analysis shows that that does not. With this, one can perform a counting of states. Essentially one has to solve a combinatorial problem of two sequences of half integers J_I, m_I, with the constraints discussed above such that the area has eigenvalues in the range $[a_0 - \delta, a_0 + \delta]$. This problem has been neatly solved in Agulló *et al.* (2010),

$$S(a) = \frac{a}{4\ell_{\text{Planck}}^2} - \frac{3}{2} \log \left(\frac{a}{\ell_{\text{Planck}}^2} \right) + O(1) + \cdots \tag{10.12}$$

and the expansion is because the calculations are done in the limit of large areas. As we see, the result reproduces the formula of Bekenstein. The logarithmic term was first derived by Kaul and Majumdar (2000). To get this result one has to assume that the Barbero–Immirzi parameter takes the value $\beta = 0.274067$, which results from solving an equality (Meissner 2004),

$$1 = \sum_{k=1}^{\infty} (k + 1) \exp \left(-\frac{1}{2} \beta \sqrt{k(k + 2)} \right). \tag{10.13}$$

So this would fix the value of the Barbero–Immirzi parameter for loop quantum gravity. Presumably other physical calculations whose results depend on the parameter should yield the same value. At the moment we do not know of any other physical prediction that depends on the value of the parameters so this is yet to be seen. A satisfying feature is that the calculation of the entropy has been worked out for a number of different types of black holes (with charge, with additional matter fields, with a cosmological constant, with rotation and distortion, see Corichi (2009) for references) and in all cases the same value of the Barbero–Immirzi parameter is needed to reproduce Bekenstein's formula. More recent papers have actually performed efficient numerical calculations of the number of states for a given area that do not depend on the large area approximation and discovered interesting structures in the relation of entropy and area that at the moment are not well understood (Corichi *et al.* 2007).

10.2   The master constraint and uniform discretizations

10.2.1   The master constraint program

As we have discussed, implementing the algebra of constraints of general relativity at a quantum level remains a significant challenge. In loop quantum gravity the diffeomorphism constraint is not implemented as an operator and therefore there is no chance of reproducing at a quantum level the classical algebra of constraints. Finite diffeomorphisms are well defined on the space of spin networks endowed with the Ashtekar–Lewandowski measure, but the infinitesimal diffeomorphisms generated by the constraint are not. This removes an important check on the sanity of the constructed quantum theory. Coupled with the lack of Dirac observables for general relativity in vacuum and the ensuing difficulties of defining a semiclassical limit, they remain the main stumbling blocks to progress in the full theory.

The master constraint program, pioneered by Thomas Thiemann and collaborators, seeks to improve the situation by substituting the constraints by a single *master constraint* which is essentially the spatial integral of the square of the constraints in question, appropriately densitized. So, for instance, taking the Hamiltonian constraint we considered in Chapter 8, in its un-smeared version,

$$\tilde{H}(x) = \left\{ A_c^k, V \right\} F_{ab}^k \epsilon^{abc}, \tag{10.14}$$

the master constraint would be defined as,

$$M = \frac{1}{2} \int d^3x \frac{\tilde{H}^2(x)}{\sqrt{\det(q)}}. \tag{10.15}$$

Notice that this is only one constraint whereas $\tilde{H}(x)$ were infinitely many. It is clear that if M vanishes so do the infinitely many $\tilde{H}(x)$'s. One may ask if it is legitimate to claim that the two pictures are equivalent, at least at the classical level. For instance, consider the Poisson bracket of the master constraint with any quantity. Because the master constraint is quadratic in the constraint, when you compute its Poisson bracket with any quantity, the result is proportional to the constraint, therefore it vanishes when constraints are enforced. So it seems that the notion of observable is lost. But if you consider

$$\{\{M,O\},O\} = 0 \qquad (10.16)$$

this condition is equivalent to O being a Dirac observable. So the master constraint can capture the information about observables.

The master constraint is diffeomorphism invariant (and $su(2)$ invariant as well). And being a single constraint it commutes with itself. So if one considers the master constraint together with the diffeomorphism constraint, they have a very simple constraint algebra,

$$\{C(\vec{N}), M\} = 0, \qquad (10.17)$$

$$\{M, M\} = 0 \qquad (10.18)$$

and the usual algebra between diffeomorphisms. This is a huge advantage at the time of quantization. The task is to promote the master constraint to a quantum operator and to find the quantum states that are annihilated by it. The advantage is that because it is a diffeomorphism invariant quantity, there is no doubt that it can be promoted to an operator on the space of diffeomorphism invariant states. And the issue of the structure functions in the algebra of constraints is bypassed. The resulting quantization will not necessarily be equivalent to a canonical quantization in all cases. So this can be seen as a generalization of Dirac's canonical quantization procedure.

The only caveat is, what happens if one discovers that as a quantum operator the master constraint does not have zero among its eigenvalues? In that case the proposal is to consider the smallest eigenvalue. One would not be dealing with a theory where the constraints are enforced exactly but with a theory where the constraints are small. Therefore the theory will not have the same exact same symmetries as the classical theory one started with but will have symmetries that approximate those of the classical theory. On the other hand, getting zero as an eigenvalue for the master constraint will be a guideline to deal with the types of ambiguities we mentioned when discussing the Hamiltonian constraint.

We will not discuss in detail the quantization of the master constraint since it proceeds along lines very similar to the ones of the Hamiltonian constraint of Chapter 8. Details can be seen in Thiemann (2006). But suppose one manages to find a self adjoint operator $\hat{M}$. Then there is a way to construct quantum Dirac observables as follows. You first construct the unitary one-parameter family of operators $\hat{U}(t) = \exp\left(it\hat{M}\right)$.

Then consider an operator $\hat{O}$ that is invariant under spatial diffeomorphisms. From there one constructs, if the limit exists, the "time averaged operator," $\widehat{\overline{O}}$,

$$\widehat{\overline{O}} = \lim_{T \to \infty} \frac{1}{2T} \int_{-T}^{T} dt \hat{U}(t) \hat{O} \hat{U}(t)^{-1}. \tag{10.19}$$

What one is doing here is integrating over the "time evolution" generated by the master constraint. It is clear that the resulting operator must commute with the master constraint itself, since commuting is equivalent to an infinitesimal time shift and one has summed over all such shifts. It is equivalent to having the real line from minus infinity to infinity and "shifting it a bit to the right;" it obviously does not change. So there exists a procedure, albeit somewhat formal, to construct Dirac observables using the master constraint.

The master constraint program has been satisfactorily tested in numerous model systems including mechanical systems, systems with challenging constraint algebras, free and interacting field theories. See Dittrich and Thiemann (2006) and references therein for further details. The master constraint program has evolved into a fully combinatorial treatment of gravity known as Algebraic Quantum Gravity, but this exceeds the scope of this book, we refer the reader to Giesel and Thiemann (2007) for further details.

10.2.2 Uniform discretizations

The technique of discretizing a field theory in order to quantize it has been very successful in the context of Yang–Mills theories. It is known as lattice gauge theory and it is a technique especially suited for treatments on the computer. Unfortunately, when one discretizes theories with symmetries it is delicate work to keep the symmetries intact. This is particularly the case in general relativity. If one discretizes spacetime, diffeomorphism invariance is immediately lost. At the canonical level this has the following implication: the equations that used to be constraints, relating variables at a given instant of time and that are automatically preserved by the evolution equations, are not preserved anymore. To enforce simultaneously the constraints and equations of motion becomes impossible. The only way to enforce all equations at the same time is to fix the value of the Lagrange multipliers. But then they are not free anymore and the character of the theory has changed dramatically. Worse, the equations that determine the Lagrange multipliers are algebraic equations (when one discretizes all derivatives disappear being replaced by finite differences). The algebraic equations are non-linear and therefore generically may involve complex solutions. Of course once the Lagrange multipliers become complex numbers, one is not approximating the continuum theory one started with, which was formulated in terms of real variables. Moreover, the Lagrange multipliers, in particular the lapse, are a measure of the rate at which one evolves. If the equations of the theory dictate that the lapse is large then it means that in the discrete time evolution there will be large jumps between successive points and one does not have a way of controlling how well the discrete theory approximates the continuum theory. These problems have been a genuine impediment in developing a lattice gauge theory version of Hamiltonian general relativity.

When one discretizes an action there is a large amount of ambiguity in how to discretize. Derivative operators can be approximated by finite differences in a large variety of ways. For instance one can compute the derivative of a function at a point by considering a finite difference between the value of the function at the point and the neighboring point to the right. Or to the left. Or to consider a symmetric difference between the values at the left and the right of the point of interest, just to name a few examples. The *uniform discretization* approach takes advantage of this freedom in the discretization of a theory to cast the evolution equations into a form in which evolution is generated by the master constraint,

$$A_{n+1} = A_n + \{A_n, M\} + \{\{A_n, M\}, M\} + \cdots \tag{10.20}$$

where A is any of the canonical variables of the theory. The beauty of this particular form of the evolution equations is that the value of the master constraint M is preserved exactly. So if one starts with a small value (meaning that one is close to the continuum theory where the master constraint vanishes), one remains close to the continuum theory upon evolution. Suppose we choose that small value to be $\delta^2/2$ and let us say we are dealing with a theory with N constraints $\phi^i(q, p) = 0$. If you define $\lambda_i = \phi_i/\delta$ (which means $\sum_{i=1}^{N} \lambda_i^2 = 1$) then the evolution of one of the dynamical variables, say, q can be expanded in δ and one gets,

$$q_{n+1} = q_n + \sum_{i=1}^{N} \{q_n, \phi_i\} \lambda_i \delta + O(\delta^2) \qquad (10.21)$$

and we recognize in the second term the usual evolution one would get with a total Hamiltonian $H_T = \sum_{i=1}^{N} \lambda_i \phi_i$. So we are getting to leading order, the traditional evolution equation for a totally constrained system like the ones we discussed in Chapter 4, only discretized. The "step" in the evolution is controlled by the value of δ and we choose that value by picking initial data such that the master constraint evaluated on them is $\delta^2/2$. So we see we have complete control over the approximation.

In the theory we are considering the constraints are not zero. Therefore the quantization of these theories proceeds as if they were unconstrained. In particular one does not have to worry about the constraint algebra. What one needs to worry about is that the discrete quantum theory should have a master constraint that contains zero among its eigenvalues. If that is the case, that would correspond to the continuum limit (a zero master constraint means a vanishing step δ in the evolution). If the master constraint does not have zero among its eigenvalues but its smallest eigenvalue is small (in units of $\hbar$) then one is left with a discrete theory that approximates the continuum theory very well.

Uniform discretizations have been tested in similar models to the ones in which the master constraint program has been tested (Campiglia *et al.* 2006). It is also being applied in spherically symmetric quantum gravity coupled to scalar fields (Gambini *et al.* 2009b).

10.3 Spin foams

10.3.1 Path integrals in quantum mechanics

In addition to the Schrödinger and Heisenberg formulations of quantum mechanics there exists another formulation called the *path integral formulation*. In the Schrödinger picture the central role is played by states, in the Heisenberg representation it is operators, and in the path integral formulation it is the transition probabilities which occupy center stage. As an example, consider a single particle located at point x_i at time t_i, and we wish to find the probability that it will be located at x_f at time t_f. The probability is calculated as follows: first enumerate all the classical paths from the initial to the final position, then calculate the classical action evaluated at each path. Assign each path a "transition amplitude" proportional to $\exp(iS/\hbar)$, the proportionality constant chosen to assure normalization. Then sum the amplitude over all possible paths. Usually this sum is an integral since there exists a continuum of paths, this is the *path integral*. The resulting sum is the transition amplitude and its square is the transition probability. For different problems, such as a particle changing from one momentum to another or for an initial state without definite position or momentum, variations of the approach apply.

The path integral (also known as "sum over histories") formulation, originally developed by Feynman, is rarely the most straightforward way to approach problems in elementary quantum mechanics. There are, however, many contexts in which it is highly desirable, like in quantum field theory, particle physics, and in general in many particle systems as in condensed matter physics. For instance, coupled with the powerful Monte Carlo statistical method to evaluate the integral, it is widely used in computations for complex systems that are carried out numerically. Some people also prefer the formulation because it deals directly with something observable, that is the transition probabilities rather than with the unobservable wavefunctions of the other formulations.

Let us motivate the path integral a bit. Consider a simple mechanical system with time independent Hamiltonian. In such a system the quantum states evolve as,

$$|\psi(t)\rangle = \exp\left(-\frac{i}{\hbar}Ht\right)|\psi(0)\rangle. \qquad (10.22)$$

If we write this expression in the position representation and using the completeness of the position basis, we get,

$$\langle x|\psi(t)\rangle = \int dx' \langle x|\exp\left(-\frac{i}{\hbar}Ht\right)|x'\rangle\langle x'|\psi(0)\rangle, \qquad (10.23)$$

with $P(x,t,x',0) = \langle x|\exp\left(-\frac{i}{\hbar}Ht\right)|x'\rangle$ the propagator or transition amplitude from $x',0$ to x,t. As an example, if we consider the free particle, which has energy eigenstates given by

$$\psi_p(x) = \frac{1}{\sqrt{2\pi\hbar}}\exp\left(\frac{i}{\hbar}px\right), \qquad (10.24)$$

one can compute the propagator as,

$$P(x, t, x', 0) = \int dp \, \langle x, t | p \rangle \langle p | x', 0 \rangle \tag{10.25}$$

$$= \frac{1}{2\pi\hbar} \int dp \exp \left(-\frac{i}{\hbar} \frac{p^2}{2m} t \right) \exp \left(\frac{i}{\hbar} p(x' - x) \right) \tag{10.26}$$

$$= \sqrt{\frac{m}{2\pi i \hbar t}} \exp \left(\frac{i}{\hbar} \frac{m(x' - x)^2}{2t} \right). \tag{10.27}$$

This expression inspired Dirac, who noted that given the classical trajectory of a free particle $x(s) = x + (x' - x)\frac{s}{t}$ if one computed the classical action evaluated on the trajectory, one gets

$$S = \frac{m}{2} \int_0^t ds \, \dot{x}^2 = \frac{m}{2} \frac{(x' - x)^2}{t}, \tag{10.28}$$

which indicates that the propagator goes as $\exp(iS/\hbar)$. Dirac did not develop the formalism further and at this stage one could think that this is just a coincidence given the simple action for a free particle. To see that this is not the case, let us consider a particle in a potential. The total energy is $E = T + V$ the kinetic plus potential energy. Let us partition the evolution from 0 to t in intervals of length $\Delta t = t/N$ and we will eventually take the limit $N \to \infty$. The evolution operator will be given by $\exp(-Et/\hbar) = [\exp(-E\Delta t/\hbar)]^N$. We would like to separate this expression into terms involving the kinetic and potential terms separately. To do so we observe that the Baker–Campbell–Hausdorff formula implies,

$$\exp \left(-\frac{i}{\hbar}(T + V)\Delta t \right) = \exp \left(-\frac{i}{\hbar} T \Delta t \right) \exp \left(-\frac{i}{\hbar} V \Delta t \right) +$$

$$\frac{1}{\hbar^2}[T, V](\Delta t)^2 + \ldots \tag{10.29}$$

and since we are interested in the limit $\Delta t \to 0$ we will keep only the first term for the "short time" evolution operator,

$$U(\Delta t) = \exp \left(-\frac{i}{\hbar} T \Delta t \right) \exp \left(-\frac{i}{\hbar} V \Delta t \right), \tag{10.30}$$

in terms of which we can write the propagator, again using the completeness of the position basis, as,

$$P(x', t, x, 0) = \lim_{N \to \infty} \int_{-\infty}^{\infty} \left(\prod_{j=1}^{N-1} dx_j \right) \langle x'|U(\Delta t)|x_{N-1}\rangle \dots \times \langle x_1|U(\Delta t)|x\rangle \dots$$

$$(10.31)$$

Each factor in the above expression can be computed,

$$\langle x_{m+1}|U(\Delta t)|x_m\rangle = \langle x_{m+1}| \exp\left(-\frac{i}{\hbar} T \Delta t\right) |x_m\rangle \exp\left(-\frac{i}{\hbar} V(x_m) \Delta t\right)$$

$$(10.32)$$

$$= \sqrt{\frac{m}{2\pi i \hbar \Delta t}} \exp\left[\frac{i}{\hbar}\left(\frac{m}{2}\frac{(x_{m+1} - x_m)^2}{\Delta t} - V(x_m) \Delta t\right)\right]$$

where in the first identity we used the fact that the potential is diagonal in the position representation, and in the second identity we used the result for the propagator of the free particle we derived above. Putting them all together, we get for the transition amplitude,

$$P(x_N, t, x_0, 0) = \lim_{N \to \infty} \int_{-\infty}^{\infty} \left(\prod_{k=1}^{N-1} dx_k \sqrt{\frac{m}{2\pi i \hbar \Delta t}} \right)$$

$$\times \exp\left[\frac{i}{\hbar} \sum_{j=0}^{N-1} \left(\frac{m}{2}\left(\frac{x_{j+1} - x_j}{\Delta t}\right)^2 - V(x_j)\right) \Delta t\right], \quad (10.33)$$

and we see that in the exponent a discretized version of the classical action appears $S = \int dt(m\dot{x}^2/2 - V(x))$. This leads us to write,

$$P(x_N, t, x_0, 0) = \int \mathcal{D}x \exp\left(\frac{i}{\hbar} S[x]\right), \qquad (10.34)$$

and this is what is usually known as *path integral* and $\mathcal{D}x$ is the measure of integration, that comparison with (10.33) shows is non-trivial. The integral in (10.34) is a functional integral in that one has to integrate over all trajectories satisfying the boundary conditions. Operationally this is what is done via the discretization in (10.33).

10.3.2   Path integrals in general relativity and spin foams

In general relativity the path integral is about the transition between geometries. Given an initial geometry on a three-dimensional slice of space-time and a final one, what is the probability of a transition? Formally one can write something like this,

$$P\left((A_a^i)_{\text{final}}, (A_a^i)_{\text{initial}}\right) = \int \mathcal{D}N\mathcal{D}N^a\mathcal{D}\lambda^i\mathcal{D}\tilde{E}\mathcal{D}A \exp(-\frac{i}{\hbar}S(\tilde{E}, A))$$

(10.35)

where $S(\tilde{E}, A)$ is the action for general relativity, which can be written as $\int d^4x \tilde{E}_i^a \dot{A}_a^i - NC - N^aC_a - \lambda^i\mathcal{G}^i$ where N is the lapse, N^a the shift, λ^i the Lagrange multiplier associated with Gauss' law $\mathcal{G}^i$, C is the Hamiltonian constraint, and C_a the diffeomorphism constraint. As we mentioned in the example of the particle, these types of expressions are formal. But in the gravitational case the distance between this expression and an actual procedure to compute the integral is quite formidable. To begin with, trajectories take place in an infinite dimensional space, since one is dealing with a field theory. Moreover trajectories have to satisfy the constraints. That somehow has to be reflected in the measure of integration. We are also integrating on the Lagrange multipliers. What range of values should the integrals take? For years these difficulties hampered development of the path integral approach to quantum gravity.

With the development of the mathematical tools of loop quantum gravity, interest naturally arose in the possibility of using some of these techniques to define the path integral. The result is the *spin foam* approach to quantum gravity, pioneered by Reisenberger and Rovelli (1997), that has by now received contributions from many authors. A spin foam is a geometric structure built from vertices, edges, and polygonal faces[1]. The faces are labeled by a group representation, pretty much like the lines of a spin network. The edges are labeled by intertwining operators. This looks like a bunch of soap suds and hence the name spin foam. If one takes a slice of a spin foam the result is a spin network. Slicing a polygonal face produces a line in the spin net with an associated group representation, slicing one of the lines of the spin foam produces a point where lines meet in the spin network with the associated intertwiner. What one would like to achieve is to compute transition amplitudes between spin network states by summing over all possible spin foams going from one spin network to the other. This is not completely understood today for general relativity, though much more progress has been made in the context of a much simpler theory, called BF theory. In four dimensions a BF theory has a Lagrangian

---

[1]Such structures, obtained by gluing together points, line segments, triangles, and their higher dimensional generalizations (simplexes) are known as *simplicial complexes*.

$$L = \int d^3x \epsilon^{\mu\nu\lambda\kappa} \text{Tr}\left(B_{\mu\nu}F_{\lambda\kappa}\right), \tag{10.36}$$

where $F_{\mu\nu}$ has the same form as in a Yang–Mills theory in terms of a vector potential and B is just an antisymmetric tensor whose components are elements of an algebra. Varying the Lagrangian with respect to B one gets immediately the equation of motion $F_{\mu\nu} = 0$ which implies, coupled with the equation for B, that the vector potential differs from zero by a gauge transformation (for trivial topologies of the manifold at least). The theory, therefore, does not have local degrees of freedom. That makes it dramatically simpler than general relativity. The only degrees of freedom one can have will be related to topological aspects of the fields. Amazingly, in a certain sense general relativity is a particular case of a BF theory. If one chooses the field B to be the product of two tetrads $B_{ab}^{IJ} = E_{[a}^I E_{b]}^J$ (tetrads are similar to triads but in four dimensions with the index I running from 0 to 3), one can show that the action corresponds to general relativity. One should not be confused into thinking that because BF theory is simple and well understood this means that general relativity should be simple. The requirement that B be given by the product of two tetrads (known as the "simplicity constraint") actually complicates the theory quite a bit and very little is gained from approaching general relativity from this perspective.

If we go back for a second to the path integral we built for a particle in a potential, we needed to study the action of the Hamiltonian of the system on the initial state. Similarly, when one studies the path integral for gravity one ends up with actions of the Hamiltonian (constraint, in this case) acting on states. As we discussed in Chapter 8 the action of the Hamiltonian constraint is to add a line. If we look at Fig. 10.2 we see that in this particular spin foam a line has been added. That would correspond to the action of a Hamiltonian constraint that takes place at the vertex where the new planar surface starts. Different formulas for assigning a weight to the vertices have been proposed over the years. The first was due to Barrett and Crane (2000), but it was soon shown to be problematic. Significant progress has been made recently by Engle, Pereira, and Rovelli (2007) and Freidel and Krasnov (2008) in defining vertices with much better behavior. Applications of spin foams are starting to emerge. The Marseille group has used spin foams to compute a propagator for gravity that yields the correct Newtonian limit (Alesci *et al.* (2009) and references therein). It is worth mentioning that a majority of the activity

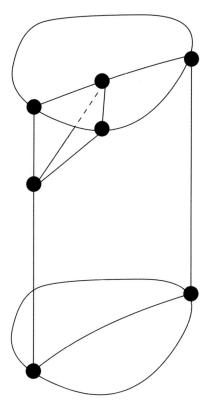

Fig. 10.2 *An example of a spin foam connecting two spin networks corresponding to the top slice and the bottom slice.*

has concentrated on the Euclidean case (they consider a four-dimensional space, rather than a space-time), as that ameliorates certain difficulties in the calculations, but also makes the connection of the results with true gravity more difficult.

To give a brief idea of some of the techniques researchers are using for constructing the vertices, it is good to introduce the concept of *Regge Calculus*. In that approach one approximates a curved space-time with flat simplexes, much like the triangles of a geodesic dome approximate a two-dimensional sphere. The idea is that the curvature in both cases is concentrated at the vertices where the triangles meet (in three dimensions curvature is concentrated along lines where tetrahedra meet, and for higher dimensions it is concentrated in (hyper)surfaces). Suppose you actually wanted to build a geodesic dome. If you specified the length of all the sides of the triangles involved, that would be enough information for an engineer to build the dome. Similarly, giving the length of the sides of all the simplexes is equivalent to completely specifying the manifold, including the data about its curvature. This key idea was used by Tullio Regge (1961) to write an approximation to the action of general relativity that was a function of the lengths of the links on a simplex $S = S(l_1, l_2, \ldots, l_N) = \sum_v S_v$ where the sum is over all simplexes and S_V is an expression dependent on the lengths of the links of a given simplex. The equations of motion can be worked out and they have even been used in numerical simulations in situations of interest. From the point of view of the path integral, one could now envision computing $\int dl_1 dl_2 \ldots dl_N \prod_v e^{iS_v(l_1,\ldots,l_N)}$. Ponzano and Regge (1968) studied in some detail this expression in the case of three-dimensional gravity in the ansatz that the lengths $l_n = j_n/(2\ell_{\text{Planck}})$ are half-integer multiples of Planck length, and they showed that in the limit of large j_n's the expression for the action is related to the Wigner $6j$ symbol. Nowadays researchers are trying to carry out similar analyses in four dimensions using different forms of the action but attempting also to compute a "vertex" just like Regge and Ponzano did in three dimensions.

10.4 Possible observational effects?

As we stated in the introduction, there does not exist a definitive experiment that requires quantum gravity for its understanding at the moment. In fact, finding such an experiment is considered a major milestone yet to be reached. The reason for this situation is that the natural scales that are determined by the coupling constants of general relativity and Planck's constant are far removed from what one can experimentally test. In terms of energy, the Planck energy is 10^{19}GeV, which is fifteen orders of magnitude larger than that which the most energetic particle accelerators can deliver and ten orders of magnitude away from the most energetic observed cosmic rays. Phrased in terms of length, Planck's length is 10^{-33}cm, about 20 orders of magnitude smaller than the radius of a proton. In terms of time, Planck's time is 10^{-44}s, more than 20 orders of magnitude away from accuracy of the best atomic clocks foreseeable in the next few years. Just on dimensional grounds only, one immediately sees that quantum gravity effects are far removed from the observable universe. We would like to discuss in this section two areas where such overwhelming odds might be improved a bit.

10.4.1 Quantum gravity and the arrival time of gamma rays from gamma ray bursts

Given what we just discussed, it came as a bit of a surprise when Amelino-Camelia *et al.* (1998) proposed that quantum gravity effects could potentially leave an imprint in observations of gamma ray bursts. The basic rationale is as follows: assuming that space-time has a granular structure with a typical size of Planck length, propagation of light of wavelength λ is expected, just on dimensional grounds, to be affected by quantum gravity effects at a level $\ell_{\text{Planck}}/\lambda$ per each wavelength. This effect is, for most wavelengths of interest, very tiny. For gamma rays, $\lambda = 10^{-12}m$ and so $\ell_{\text{Planck}}/\lambda \sim 10^{-23}$. However, gamma ray bursts occur at cosmological distances up to $L = 10^{10}$ in light years or 10^{25} in meters. In terms of wavelengths one is talking about 10^{37} wavelengths, plenty of opportunity to influence the propagation of gamma rays. What would one actually see? Interaction with a granular structure generically produces anomalies in the dispersion relation (the relation between frequency and wavelength, or equivalently momentum and energy). Using a model of string theory, Amelino-Camelia et al. conjectured that the dispersion relation for photons was

$$c^2 p^2 = E^2 \left[1 + \chi \frac{E}{E_{\text{Planck}}} + O\left(\frac{E^2}{E_{\text{Planck}}^2} \right) \right], \qquad (10.37)$$

with χ a number of order one. Gamma ray bursts emit gamma rays in a spectrum of energies. Rays of different energy would take different times to arrive at a detector, due to the anomalous dispersion relation. The differences in times could be detectable if indeed χ is non-zero. Several researchers have expressed skepticism about having a non-zero χ, since if one takes square roots of both members of the dispersion relation that implies a term involving a fractional power of the Planck energy and that does not seem to be something to arise naturally. This led many to believe that such a term is indeed zero and the first order correction is the last term in the above dispersion relation. Because that brings in an extra power of the Planck energy, it renders its observation much more difficult.

Given the limitations we have discussed in making contact with physics in loop quantum gravity, a detailed calculation of the propagation of a gamma ray through a quantum space-time is beyond the realm of possibilities. Just as an exploration of what could be possible, Gambini and Pullin (1999) carried out a simple toy calculation in loop quantum gravity. The idea was to consider the Hamiltonian of Maxwell theory in a curved space-time,

$$H = \frac{1}{2} \int d^3x \frac{q_{ab}}{\sqrt{\det(q)}} \left(\tilde{e}^a \tilde{e}^b + \tilde{b}^a \tilde{b}^b \right), \tag{10.38}$$

where we have kept track that the electric field is a density, and in this context it is also natural for the magnetic field to be a density. This forced the inclusion of the determinant of the spatial metric in the denominator in order to have an integrand that is a density of weight one. The idea then is to use the same trick due to Thiemann that we discussed in the section of the Hamiltonian constraint to represent $q_{ab}/\sqrt{\det(q)}$. We also point split the electric and magnetic fields and work in an approximation in which we take the electric and magnetic fields to be classical and smooth functions of the manifold. We also take a "weave" state for the gravitational field $|\Delta\rangle$, which corresponds to a spin network with many links typically separated by distances larger than the Planck length, $|\Delta\rangle$ but much smaller than the wavelength of light considered, and is a state expected to approximate a semi-classical geometry. One then gets for the expectation value of the electric part of the Maxwell Hamiltonian,

$$\langle \Delta | \hat{H}^E_{\text{Maxwell}} | \Delta \rangle = \sum_{v_i, v_j} \langle \Delta | \hat{w}_a(v_i) \hat{w}_b(v_j) | \Delta \rangle e^a(v_i) e^b(v_i), \tag{10.39}$$

where we have used the fact that the Thiemann trick picks up contributions only at vertices v_i, approximated the integral as a sum over all vertices of the spin network, translated the point splitting into considering adjacent vertices v_i, v_j, and represented the metric operator as a point-split product of two operators $\hat{w}_a$. We then use the fact that at the level of each "elementary cell" of the spin network weave, the electric and magnetic field are very smooth functions (their wavelength is many orders of magnitude larger than the typical cell size) to expand,

$$\tilde{e}^a(v_i) = \tilde{e}^a(P) + (v_i - P)_c \partial^c \tilde{e}^a(P) + \dots \tag{10.40}$$

where P is the central point of the cell. Notice that $(v_i - P)^c$ is a vector of magnitude Δ whereas the derivative of the field is of order of the wavelength λ, this is an expansion in Δ/λ. Inserting this expansion in the

Hamiltonian, the leading term is just the classical Maxwell Hamiltonian. Let us consider the next term,

$$\frac{1}{2} \sum_{v_i, v_j} \langle \Delta | \hat{w}_a(v_i) \hat{w}_a(v_j) | \Delta \rangle \left((v_i - P)_c \, \partial^c \left(e^a(P) \right) e^b(P) + \right.$$

$$\left. (v_j - P)_c \, \tilde{e}^a(P) \partial^c \left(\tilde{e}^b(P) \right) \right). \tag{10.41}$$

When performing the sum over all the cells in the weave one needs to evaluate the quantity,

$$\langle \Delta | \hat{w}_a(v_i) \hat{w}_b(v_j) | \Delta \rangle \frac{(v_i - P)_c}{\Delta}. \tag{10.42}$$

Such a quantity averages out to zero in a first approximation since one is summing a vector over a set of vertices that one assumes to be isotropic on average. The value of this quantity is therefore at best proportional to $\ell_{\text{Planck}}/\Delta$ or higher orders of such quantity. Here we assume it is of first order to study the implications of such a situation. The form of this quantity has to be a rotationally symmetric tensor, that is, it is given by $\chi \epsilon_{abc} \ell_{\text{Planck}}/\Delta$ with χ a factor of order one. We therefore have a corrective term to the Maxwell Hamiltonian. The term is rotationally invariant but it violates parity. To see this, one works out the corrected Maxwell equations (we go back to the usual vector notation with $\vec{E}$ and $\vec{B}$ the electric and magnetic fields for clarity),

$$\partial_t \vec{E} = -\nabla \times \vec{B} + 2\chi \ell_{\text{Planck}} \left(\nabla^2 \vec{B} \right), \tag{10.43}$$

$$\partial_t \vec{B} = \nabla \times \vec{E} - 2\chi \ell_{\text{Planck}} \left(\nabla^2 \vec{E} \right), \tag{10.44}$$

and the corrective terms depend on the flat space Laplacian of the fields. Notice that this set of equations is not Lorentz invariant, the weave selects a preferred frame. If one combines the equation to get a wave equation as usual, one gets,

$$\partial_t^2 \vec{E} - \nabla^2 \vec{E} - 4\chi \ell_{\text{Planck}} \nabla^2 \left(\nabla \times \vec{E} \right) = 0, \tag{10.45}$$

and a similar equation for $\vec{B}$. If one seeks solutions of these equations given by a plane wave in the $\vec{k}$ direction with a given helicity,

$$\vec{E}_\pm = \text{Re} \left((\vec{e}_1 \pm i\vec{e}_2) \, e^{i\left(\Omega_\pm t - \vec{k}\cdot\vec{x}\right)} \right), \tag{10.46}$$

where $\vec{e}_{1,2}$ are a basis of vectors in the directions perpendicular to the direction of wave propagation k, and Ω is the frequency of the wave, one finds that such waves solve the equations provided that,

$$\Omega_{\pm} = \sqrt{k^2 \mp 4\chi\ell_{\text{Planck}}k^3} \sim |k|\left(1 \mp 2\chi\ell_{\text{Planck}}|k|\right) + \ldots \qquad (10.47)$$

and this is the dispersion relation of the wave. As we see one gets the usual leading term of $\Omega_{\pm} \sim |k|$ for ordinary photons but one gets a correction that is a different sign for different helicities. Quantum space-time is therefore *birefringent*. Being able to tell right-handed waves from left-handed waves is evidence of the loss of parity invariance. This effect is different from the one found by Amelino-Camelia *et al.* in string theory. There they did not find helicity dependence but just corrections to the dispersion relation proportional to higher powers of $|k|$. Such corrections also appear here, but at higher orders in the expansion in terms of $\ell_{\text{Planck}}/\lambda$.

It seems impossible to over-stress the severe limitations of this calculation. First of all, the constraints are not being enforced in any way, it is a kinematical calculation. Nowhere have we input into the calculation that we are working in general relativity, for example. Moreover, in order to obtain the effect we had to assume that the quantum state was a weave that violates parity. There is no physical motivation to argue that such a state should describe our universe. Therefore at best the calculation is a reflection of an effect that could be *possible* in loop quantum gravity but by no means certain.

It turns out that the effect is actually already ruled out, not by gamma ray bursts but by radioastronomy. There exist radio sources which produce polarized waves in a wide range of wavelengths and the polarization does not change with wavelength. Gleiser and Kozameh (2001) estimated that this constrains the parameter χ to values less than 10^{-3}. A more severe bound can be placed using certain gamma ray bursts (Mitrofanov 2003). The effect proposed by Amelino-Camelia *et al.* (10.37), which does not involve birefringence, is less severely constrained (Abdo *et al.* 2009).

Unfortunately some authors and bloggers have mistaken the experimental bound of the effect as "ruling out loop quantum gravity" (For example Mitrofanov 2003). It should be clear from the above discussion that the experimental constraint is at best a limitation on the type of state that may hypothetically describe our universe and is in no way ruling out a definite prediction of the theory. It is only a hypothetical state since we have not enforced at all the dynamics of the theory or in any way suggested a mechanism that may predict that such a state is likely to occur in nature. A true prediction that could rule out the theory if it contradicted an experiment is something that the current state of the art of the theory cannot generate.

There are larger conceptual questions that are brought up by the above calculation that are not satisfactorily understood to late. The most obvious is that the effective corrected Maxwell equations that arise are not Lorentz invariant. Breaking Lorentz invariance is a very dangerous thing for a theory since it can quickly lead to large experimental discrepancies (Collins *et al.* 2004).

10.4.2 Limitations in measurements of times and lengths

There have been suggestions in the literature that the characteristic scale at which quantum gravity effects occur could potentially be significantly larger than the Planck scale. There has only been very marginal evidence of such claims within loop quantum gravity (Gambini and Pullin (2008a)). Nevertheless these suggestions have been the source of speculation and inspiration to keep on looking for possible effects of quantum gravity, since they suggest effects culd occur at energies and scales commensurate with modern precision experiments.

The argument goes along the following lines. Some time ago, Salecker and Wigner (1958) studied what limitations quantum mechanics placed on the measurement of space-time. In particular they asked the question: what is the ultimate accuracy a clock can achieve? To probe this they imagined an ideal clock consisting of two mirrors with a photon bouncing between them. Every time the photon hits a mirror the clock "ticks". To try to eliminate all outside interferences with the experiment, they assume the clock is in a perfect vacuum and not interacting with any external disturbances. Even such an idealized clock develops inaccuracies, mainly due to the fact that the wavefunction of the mirrors, which are not interacting with an environment, spread. As they spread, the moment in which the "tick" occurs becomes less accurate. Salecker and Wigner estimate such accuracy as $\delta t \sim \sqrt{t/M}$ where t is the time one wishes to measure. The formula is in units in which $\hbar = c = 1$ but G is not one. In these units mass has dimensions of inverse time. As one sees, the inaccuracy of the clock grows with the time one wishes to measure, as one would expect. It goes as the inverse of the mass of the mirrors. The message one gets is: if one wishes for a more accurate clock, one needs to make it more massive.

But this argument does not take gravity into account. Gravity actually precludes a clock from being infinitely massive, long before that it would have turned into a black hole. You might ask, can one make the clock more massive and at the same time bigger, and therefore avoid the high densities needed to form a black hole? The answer is negative: bigger clocks become more inaccurate due to speed of light limitations. This argument has been put forward by many authors going back to Karolhyazy, *et al.* 1986 and more recently by Ng and van Dam (1995) and others. This creates a tension with the Salecker–Wigner argument: it is not true anymore that making things more massive improves accuracy. The best accuracy in fact is achieved for a clock with a mass just slightly under what is needed to form a black hole for a given clock size. This gives an accuracy of $\delta t \sim t_{\text{Planck}}^{2/3} t^{1/3}$ where $t_{\text{Planck}} \sim 10^{-44} s$. To better understand the uncertainty it is worth rewriting it as,

$$\delta t \sim \sqrt[3]{\frac{t}{t_{\text{Planck}}}} t_{\text{Planck}}. \tag{10.48}$$

So the effect is "of the order of Planck time" as one would expect a time effect arising from quantum mechanics and gravity, but the factor in front of the Planck time is large. For t of the order of what one may consider in a lab (days, hours), the factor is of the order of 10^{15}. That is still quite small, it is about ten orders of magnitude smaller than the accuracy of the best atomic clock, but it raises hopes that perhaps some clever arrangement could allow us to detect it. For instance, such inaccuracy in time leads to an inaccuracy in the measurements of distances of 10^{-18} meters. Gravitational wave interferometers like the LIGO and VIRGO projects are able to probe distances of that order. This does not immediately mean that the effect could be observable, interferometers are complex instruments that measure average positions using many photons and smear away any uncertainty in the arrival of individual photons. But the fact that the orders of magnitude are comparable is suggestive.

Notice that the arguments presented are handwaving in nature. They are not really based on quantum gravity since in the argument gravity is entered as a classical concept. Right now the most favorable light in which these arguments can be cast is to give them the same status that the uncertainty principle had in the early 1920s. Then people were aware that such a principle could exist but they did not have a detailed understanding of it nor a fundamental reason to think it should always hold. In the end such arguments were found. Whether a similar fundamental underlying argument can be found from a quantum theory of gravity for these types of uncertainties is yet to be seen. The arguments have been criticized in the literature (Requardt 2008), but it should be noted that similar bounds can be found with other arguments (Lloyd and Ng (2004)).

10.5   The problem of time

As we have discussed, general relativity is a totally constrained theory. The total Hamiltonian is just given by a linear combination of constraints. Since the only quantities one can physically study in a constrained theory are quantities that have vanishing Poisson brackets with the constraints (they are invariant under the symmetries of the theory), this implies in general relativity that such physical constants do not evolve. One is left with a *frozen formalism* and one is supposed to somehow disentangle the dynamics from it. Some have considered this as the central conceptual obstacle to the canonical quantization of gravity. To quote Kuchař (1992) "In my opinion, none of us has so far succeeded in proposing an inter- pretation of quantum gravity that would either solve or circumvent the problems of time." In this section we would like to discuss some recent proposals that have been made to deal with this issue.

10.5.1   Evolving constants of the motion

This proposal has a long history, going back to Wheeler and Bergmann in the 1960s in the canonical context and to Einstein himself as an abstract idea. In later years, it has been emphasized and studied in some detail by Rovelli (1991). Given a totally constrained system with canonical variables q^i, p_i, one constructs one parameter families of Dirac observables $Q^i(t)$. These quantities have vanishing Poisson brackets with the constraints. They are such that when the parameter takes the value of one of the canonical variables, for instance $t = q^1$, then the observable $Q^i(t = q^1) = q^i$. The underlying idea is that in a theory with symmetries and constraints one cannot ask for the value of a given variable, but one can ask relational questions between the values of the variables and such questions are invariant.

Let us see how this works in a simple example, the relativistic particle in one dimension. The canonical coordinates are p_0, p, q^0, q. The constraint in question tells us that the of the four-momentum is the mass of the particle squared,

$$\phi = p_0^2 - p^2 - m^2 = 0. \tag{10.49}$$

One can immediately see that p is a Dirac observable. A little bit of reflection shows that there is another, independent, Dirac observable,

$$X = q - \frac{p}{\sqrt{p^2 + m^2}} q^0. \tag{10.50}$$

In terms of these two Dirac observables one can build an evolving constant of the motion,

$$Q(t, q^0, q, p) = X + \frac{p}{\sqrt{p^2 + m^2}} t, \tag{10.51}$$

and one can easily see that $Q(t = q^0) = q$. One therefore has an evolution in terms of the parameter t.

There are two critiques that can be made to this approach, and they relate to the role of the parameter t. To begin with, what is the physical role of the parameter t? Why will the evolution in terms of it correspond to something we observe in nature? In particular, since t is being identified with a canonical variable, if one is dealing with a constrained theory it is not one of the quantities that are physically observable. Should one not base the description of time on a variable that is physically observable? The second problem arises when one goes to the quantum theory. There, what is one to make of the requirement $t = q^0$ when q^0 is now a canonical variable that is therefore promoted to an operator whereas t remains a classical continuous parameter? Is one to choose one of the canonical variables and not quantize it, in order to identify it with a continuous parameter? We will see that there is a way of using evolving observables without having to observe the classical parameter t, but for that we need to briefly discuss another proposal that has been made for dealing with the problem of time.

10.5.2   The conditional probabilities interpretation

Page and Wootters (1983) made a proposal to address the problem of time in systems like general relativity called the conditional probabilities interpretation. It consisted in promoting all variables to quantum operators, choosing one of them as a clock and asking conditional probability questions with respect to the other variables. So suppose one wishes to study a quantum variable X and its time evolution, and the clock variable is called T and both variables commute. Both are represented by quantum operators. One could ask "what is the probability that X take a given value X_0 conditional on the time being T_0?" Such a question can be asked quantum mechanically as,

$$P(X = X_0 | T = T_0) = \frac{\langle \Psi | P_{T_0} P_{X_0} | \Psi \rangle}{\langle \Psi | P_{T_0} | \Psi \rangle}, \tag{10.52}$$

where P_{T_0} is the projector onto the eigenspace associated with the eigenvalue T_0 and similarly P_{X_0}. Ψ is the quantum state. The main problem with this approach in totally constrained theories is what does one choose as X and T? If one chooses Dirac observables then one runs into trouble since they are all constants of the motion and none of them can be a good clock T. Page and Wootters try to circumvent this problem by choosing X and T to be quantities that do not have vanishing Poisson brackets with the constraints. This could be reasonable since, although such variables are not physically observable, relations between them could have invariant meaning as we have mentioned before. But there are further problems: what does one choose for the state Ψ? Does one choose a state that is annihilated by the constraints? Or does one choose a kinematical state not annihilated by the constraints? If one chooses the latter one faces the difficulty of explaining where one enters information about what theory one is dealing with since one is not using the constraints at all in the construction. Page and Wootters propose to use states Ψ that are annihilated by the constraints. But acting on such states the projectors on the eigenstates of variables that are not Dirac observables will in general map out of the states that annihilate the constraints. Kuchař (1992) applied the construction in a simple example of a parametrized particle and showed that one obtained incorrect propagators due to the mentioned problem. Essentially the propagator comes out to be,

$$\langle t, x | t', x' \rangle = \delta(t - t')\delta(x - x'), \tag{10.53}$$

so the particle does not propagate!

10.5.3 Conditional probabilities with evolving constants of the motion

It was recently shown (Gambini *et al.* 2009a) that if one combines the previous two proposals, one can overcome several of the difficulties they encounter individually. The idea is to construct a conditional probability interpretation but where the quantities one computes conditional probabilities of are evolving constants of the motion. So one starts with a canonical system with constraints, one constructs an evolving constant of the motion corresponding to a quantity one wishes to study $X(t)$, and one picks another evolving constant with vanishing Poisson bracket with $X(t)$ that will act as a clock $T(t)$. One then computes the conditional probability similarly as was done in the Page–Wootters case, but between evolving constants of the motion,

$$P\left(X = X_0 | T = T_0\right) = \lim_{\tau \to \infty} \frac{\int_{-\tau}^{\tau} dt \, \langle \Psi | P_{T_0}(t) P_{X_0}(t) | \Psi \rangle}{\int_{-\tau}^{\tau} dt \, \langle \Psi | P_{T_0}(t) | \Psi \rangle}. \qquad (10.54)$$

Because the evolving constants of the motion are genuine Dirac observables, when acting on states that are annihilated by the constraints they produce as a result states that are still annihilated by the constraints. Moreover, the unobservable classical parameter t is integrated over in the expression of the conditional probability and we therefore do not need to find a way of observing it physically. Finally, it has been shown in simple examples to yield the usual propagators provided one chooses the clock variable judiciously. One only expects to recover the usual propagators when the clock variable is "very classical," monotonous in its growth, etc.

Of course one is still far from solving the problem of time in situations of interest, since one has to construct the evolving constants of the motion and that is in general not possible in general relativity in vacuum. But approximation schemes have been proposed (Dittrich 2006) to construct evolving constants of the motion and within this context this proposal for solving the problem of time is eminently feasible.

Further reading

Most of the material in this chapter is on current research so there are very few pedagogical introductions, the only further reading one can recommend are the references cited. For black hole thermodynamics in general the book by Carroll (2003) has a concise treatment. A good introduction to path integrals in quantum mechanics is the book by Feynman and Hibbs (1965). A good, although by now somewhat dated, introduction to spin foams is given by Perez (2004), which will be updated (Perez 2011). The books by Rovelli (2007) and Thiemann (2008) have chapters on spin foams and on black hole entropy. In particular the one by Rovelli has a nice physical discussion about what the entropy of a black hole is supposed to represent. The problem of time is beautifully summarized in the review by Kuchař (1992).

11
Open issues and controversies

At first this may sound like a strange title for a chapter. Why should there be controversies in a book about science? Is science not supposed to be based on facts? This is true of established science. When one is talking about science that is incomplete, the situation changes. Loop quantum gravity is an incomplete theory. We do not know if the current formulation, including the form of the Hamiltonian constraint we discussed, really describes nature or not. Worse, there are not many things that the current state of the art of the theory allows one to compute explicitly. Confronted with this situation scientists can have honest disagreements about how promising an approach loop quantum gravity is. This is because one is speculating about results not obtained yet and different people's intuition, gut feelings, and so on, will lead them to different conclusions. It is the scientific version of the glass half empty versus the glass half full argument. In this modern day and age a lot of the disagreements about loop quantum gravity are aired in blogs and discussion groups on the Internet. In general we do not recommend people learn from the controversy there. Because blogs and Internet discussion groups are unedited opinions of people, some of which have questionable training or knowledge of the details, and are usually written relatively quickly, in many circumstances they contain highly inaccurate statements. Fortunately some of the criticisms of loop quantum gravity have been actually published in peer-reviewed journals. Nicolai, Peeters, and Zamaklar (2005) have written an article outlining some of the concerns some scientists have. Their article is a bit dated but still worth reading. Thiemann (2007) has addressed some of the criticisms. Nicolai and Peeters (2007) have an updated version of their arguments. Here we will highlight some of the concerns raised in these papers.

One of the most common objections that one hears about loop quantum gravity is already raised about its kinematical structure. As we noted, one works in a representation in which one has the Ashtekar–Lewandowski inner product and such representation has a discontinuous nature. This was even evident in the context of loop quantum cosmology: although we were dealing with a mechanical system with a finite number of degrees of freedom, one could obtain something novel and non-trivial using loop quantum gravity techniques. This was because the discontinuity in the representation allowed one to bypass the Stone–Von Neumann theorem. But what happens if one applies similar representations to more familiar physical systems, for instance, the harmonic oscillator? This was addressed in a paper by Helling and Policastro (2004) and they reach the conclusion that one does not get the correct physics from a loop-type quantization. Thiemann 2007 rebuts this, showing that the loop-type quantization can approximate arbitrarily precisely the usual results for the harmonic oscillator, though one does not reproduce the usual quantization entirely (it would constitute a non-existent limit, very much like the $\mu_0 \to 0$ limit of loop quantum cosmology). The situation for simple quantum mechanical systems has been further studied by Ashtekar, Fairhurst, and Willis (2003) and Fredenhagen and Reszewski (2006).

Some of the objections are addressed at the specific version of the Hamiltonian constraint proposed by Thiemann that we discussed in this book. An important objection is the level of ambiguity in the construction. On the one hand, there exist factor ordering ambiguities and there is also the ambiguity of what type of loop one added at the vertex in order to construct the terms that yield the curvature and the connection in Hamiltonian constraint in the limit in which the triangulation is taken to shrink. Of course, if one allows total freedom of ordering even elementary field theories are fraught with infinitely many ambiguities. The usual guideline is to use naturalness, that is not to over-complicate expressions when proceeding to quantize them. In the case of Thiemann's Hamiltonian, since the construction leads naturally to having all the triads combined into a volume operator, the only ordering ambiguity ends up being if the volume should go to the right or the left of the holonomies. If one puts the volume to the left one runs into problems. Consider a point of the triangulation that does not correspond to a vertex of the spin net but some intermediate regular point of the net. If one added the holonomies first in such a way that the resulting state has a non-coplanar four-valent intersection, and then acted with the volume, one would get a non-zero contribution for the Hamiltonian for a point where there was no intersection on the original spin net[1]. That means that when one takes the limit of the triangulation shrinking to zero the resulting operator would get infinitely many contributions and diverge. So in reality there seems to be quite a bit of rigidity in the choice of factor ordering. There is an additional ambiguity that arises because one decided to add the holonomies in the fundamental representation of $SU(2)$. One could add the line in any representation. It has been shown in three-dimensional gravity that doing so leads to spurious solutions (Perez 2006) so it is likely that such ambiguity is also restricted. The status of the ambiguity of choice of loop used to introduce the holonomies is not well studied yet, in particular it is unclear what that uncountable number of ambiguities implies at the level of solutions of the Hamiltonian constraint. Some people have raised the question: are the ambiguities in the definition of the Hamiltonian constraint similar to the infinitely many counterterms one needs to introduce in perturbation theory in the case of a non-renormalizable theory? People working in loop quantum gravity tend to believe they are not. Arguments raised indicate that the countable number of renormalization constants take continuous values therefore constituting an uncountable infinite number of ambiguities, whereas in loop quantum gravity the physical Hilbert space depends only on a discrete number of ambiguities. Moreover, in perturbative quantum gravity even introducing infinitely many counterterms one has a theory that diverges and has to be regularized and renormalized, whereas loop quantum gravity is expected to be finite, perhaps even without renormalization. Finally, in loop quantum gravity many choices of values of the ambiguous parameters yield pathological theories that are therefore ruled out whereas in perturbative quantum gravity all possible values of the counterterms are equally natural. The subject of ambiguities is currently being debated and researched within the loop quantum gravity community and remains one of the important points to be addressed by the theory. Ashtekar (2008) summarizes the situation well: "the most unsatisfactory aspect of the current status of the program is that the physical meaning and ramifications of these ambiguities are still poorly understood. One can invoke 'naturality' and 'simplicity' criterion

---

[1]The situation is actually worse, a more careful discussion shows one gets contributions at every vertex of the triangulation, even those not touched by the original spin network state.

to remove them but without a deeper understanding of their physical meaning, these criteria can be subjective and therefore not compelling." Also, it should be noted that ambiguities become a problem when one is talking about choices that yield different theories that are all compatible with experiments. Given the status of loop quantum gravity it is rather premature to claim one is in such a position. It could well be that as we progress in our understanding of the semiclassical limit many ambiguities disappear.

Another point that has been raised has to do with matter couplings. Thiemann has shown that one can apply the same techniques used to write the Hamiltonian constraint for pure gravity to the study of gravity coupled to the standard model (Fermions, scalar fields, gauge fields including supersymmetry). It appears that matter can be coupled without any restrictions and without the introduction of any anomalies or divergences. This is in marked contrast to other approaches to quantum gravity like supergravity, where the matter content is highly restricted in order to make the theory consistent. This could actually be seen as a positive aspect since it means the theory has more possibilities for incorporating matter correctly. But one also has to emphasize that loop quantum gravity does not appear to impose consistency requirements on matter couplings at the kinematical level. Things could change when one implements the dynamics and studies the semiclassical limit. It is unclear, for instance, if one actually constructs the master constraint for gravity coupled to matter, if zero will be in its spectrum or not for certain matter couplings. So at best it is premature to view this aspect as a problem for the theory.

Another point some people worry about is: where are the infinities that appear in perturbative quantum gravity reflected in loop quantum gravity? Everything appears to be finite. Unfortunately it is hard to make contact with perturbative quantum gravity starting from the full theory. So the answer to this point is not really known. Presumably one has sorted out the problem of time, either by using one of the relational formulations or by coupling the theory to matter, one will have an effective Hamiltonian (not a constraint) that generates time evolution. One could then define scattering amplitudes. To make contact with perturbation theory one would have to consider states that are peaked at a Minkowski space-time. For instance in the semiclassical treatment of Sahlmann and Thiemann (2006) counterterms correcting the cosmological constant appear. These calculations are limited at present but give some idea of how contact with perturbative results could arises.

A point that worries several people, including many researchers in the field, is the breaking of local Lorentz invariance and/or diffeomorphism symmetry. Again, there is no conclusive evidence that such symmetries are broken. But the fact that one cannot implement the constraint algebra "off shell," that is including both the diffeomorphism and Hamiltonian constraint, is no doubt troublesome. That algebra is what guarantees classically that the $3 + 1$ formalism, in spite of treating space and time in different ways, remains invariant under space-time diffeomorphisms. Along the same lines, the fact that the diffeomorphism constraint and the Hamiltonian constraint are handled in such a different fashion is also a cause of concern. This was one of the motivations for the master constraint program, since there one could include both the Hamiltonian and diffeomorphism constraints in the construction of the master constraint and therefore treat them in the same footing, and this has been achieved in the Algebraic Quantum Gravity setting (Giesel and Thiemann 2007). Even ignoring the constraint algebra one could worry about local Lorentz invariance in a theory that has a fundamental level of discreteness. In particular how does one square the existence of a fundamental length with the Lorentz–Fitzgerald contraction? This was discussed in some detail by Rovelli and Speziale 2003. It should be noted that local Lorentz invariance has to be taken quite seriously. Even minute violations of Lorentz invariance (for instance, at the Planck scale) can translate themselves into catastrophic departures from experimental results when one does calculations in perturbative quantum field theory (see Collins *et al.* 2004 for more details).

Summarizing, the incomplete state of the theory coupled with some aspects of it that differ from usual treatments of quantum field theories have led some people to worry that something could be problematic in loop quantum gravity. Up to now no concrete well defined objection can be formulated to the theory. That does not mean that honest scientists cannot remain highly skeptical of the whole approach. Ultimately it is up to the practitioners to probe situations of increasing physical complexity and to show that the theory produces sensible physical predictions. In this context it is refreshing that in loop quantum gravity there are not many things that can be tweaked to make the theory agree with experiment. The kinematical framework as we noted is highly unique. The number of space-time dimensions one works with is also fixed. One can add matter fields, but those will not affect the predictions of the purely gravitational sector of the theory which again is severely constrained. This is in contrast to string theory, where the theory has evolved considerably over time and is now perceived by some as having too much freedom to be a predictive theory. Only time and further work will tell if loop quantum gravity is viable or not. Altogether the many lessons learned about the construction of non-perturbative approaches to quantum field theories makes the effort quite worthwhile and attractive irrespective of its final success in providing a theory of quantum gravity.

References

Abdo, A. *et al.* (2009). *Nature*, **462**, 331.

Abers, E. and Lee, B. (1973). *Phys. Rep.* **9**, 1.

Agulló, I., Barbero, J. F., Borja, E., Díaz-Polo, J., and Villasenor, E. (2010). *Phys. Rev. D* **82**, 084029.

Alesci, E., Bianchi, E., and Rovelli, C. (2009). *Class. Quant. Grav.* **26**, 215001.

Amelino-Camelia, G., Ellis, J., Mavromatons, N., Nanopoulos, D. and Sarkar, S. (1998). *Nature* **393**, 763.

Arnowitt, R., Deser, S., and Misner, C. (2008). *Gen. Rel. Grav.* **40**, 1997.

Ashtekar, A. (1986). *Phys. Rev. Lett.* **57**, 2244.

Ashtekar, A. (1988) *New Perspectives in Canonical Gravity.* Bibliopolis, Naples.

Ashtekar, A. (2008). "Loop quantum gravity: Four recent advances and a dozen frequently asked questions", in H. Kleinert, R. Jantzen, R. Ruffini (eds.), *Proceedings of the Eleventh Marcel Grossmann Meeting on General Relativity.* World Scientific, Singapore.

Ashtekar, A., Baez, J., Corichi, A., and Krasnov, K. (2000). *Adv. Theo. Math. Phys.* **3**, 418.

Ashtekar, A., Fairhurst, S., and Willis, J. (2003) *Class. Quan. Grav.* **20**, 1031.

Ashtekar, A. and Krishnan, B. (2004) *Liv. Rev. Rel.* **7**, 10.

Ashtekar, A. and Lewandowski, J. (1997) *Class. Quan. Grav.* **14**, A55.

Ashtekar, A. and Wilson-Ewing (2009). *Phys. Rev. D* **79**, 083535.

Ashtekar, A., Pawlowski, T., and Singh, P., (2006) *Phys. Rev.* **D74**, 084003.

Ashtekar, A., Pawlowski, T., and Singh, P., (2007) *Phys. Rev.* **D74**, 75, 024035.

Ashtekar, A., Singh, P. (2011) "Loop cosmology" in preparation for *Class. Quan. Grav.*

Baez, J. and Muniain, J. (1994) *Gauge Fields, Knots and Gravity.* World Scientific, Singapore.

Barbero, F. (1995), *Phys. Rev.* **D51**, 5507.

Barrett, J. and Crane, L. (2000) *Class. Quan. Grav.* **17**, 3101.

Bekenstein, J. (1973) *Phys. Rev.* **D7**, 2333.

Bianchi, E. and Rovelli, C. (2010). "Why all these prejudices against a constant?," arXiv:1002.3966 [astro-ph.CO][accessed 11 March 2011].

Bojowald, M. (2000). *Class. Quan. Grav.* **17**, 1489.

Bojowald, M. (2004). *Class. Quant. Grav.* **21**, 3541.

Bojowald, M. (2008). *Liv. Rev. Rel.* **11**, 4.

Borissov, R., De Pietri, R., and Rovelli, C. (1997). *Class. Quan. Grav.* **14**, 2793.

Brown, D. (2011). In preparation.

Campiglia, M., Di Bartolo, C., Gambini, R., and Pullin, J. (2006). *Phys. Rev.* **D74**, 124012.

Carlip, S. (2008). *Class. Quant. Grav.* **25**, 154010.

Carroll, S. (2003). *Spacetime and Geometry: An Introduction to General Relativity.* Benjamin Cummings. Based on notes available online at http://preposterousuniverse.com/grnotes/[accessed 11 March 2011].

Carter, B. (1971). *Phys. Rev. Lett.* **26**, 331.

Christodoulou, D. and Ruffini, R. (1971). *Phys. Rev.* **D4**, 3552.

Collins, J., Perez, A., Sudarsky, D., Urrutia, L., and Vucetich, J. (2004). *Phys. Rev. Lett.* **93**, 191301.

Corichi, A. (2009). *Adv. Sci. Lett.* **2**, 236.

Corichi, A. Díaz-Polo, J., and Fernández-Borja, E. (2007). *Phys. Rev. Lett.* **98**, 181301.

Date, G. (2010). "Lectures on constrained systems," arXiv:1010.2062 [gr-qc][accessed 11 March 2011].

Dirac, P. (2001). *"Lectures in Quantum Mechanics"*, Dover.

Dittrich, B. (2006). *Class. Quan. Grav.* **23**, 6155.

Dittrich, B. and Thiemann, T. (2006). *Class. Quan. Grav.* **23**, 1143.

Donoghue, J (1994). *Phys. Rev.* **D50**, 3874.

Engle, J., Pereira, R., and Rovelli, C. (2007). *Phys. Rev. Lett.* **99**, 161301.

Eppley, K. and Hannah, E. (1977). *Foundations of Physics.* **7**, 51.

Feynman, R. and Hibbs, A. (1965). *'Quantum Mechanics and Path Integrals'.* McGraw–Hill.

Fleischhack, C. (2006). *Phys. Rev. Lett.* **97**, 061302.

Fredenhagen, K. and Reszewski, F. (2006). *Class. Quan. Grav.* **23**, 6577.

Freidel, L. and Krasnov, K. (2008). *Class. Quan. Grav.* **25**, 125018.

Frolov, V. and Novikov, I. (1998). *Black Hole Physics.* Kluwer, Dordrecht.

Gambini, R., Porto, R., Pullin, J., and Torterolo, S. (2009a). *Phys. Rev.* **D79**, 041501(R).

Gambini, R. and Pullin, J. (1996). *Loops, Knots, Gauge Theories and Quantum Gravity.* Cambridge University Press, Cambridge.

Gambini, R. and Pullin, J. (1999). *Phys. Rev.* **D59**, 124021.

Gambini, R. and Pullin, J. (2008a). *Int. J. Mod. Phys.* **D17**, 545.

Gambini, R. and Pullin, J. (2008b). *Phys. Rev. Lett.* **101**, 161301.

Gambini, R., Pullin, J. and Rastgoo, S. (2009b). *Class. Quan. Grav.*, **26**, 215011.

Gambini, R. and Trias, A. (1980). *Phys. Rev.* **D22**, 1380.

Gambini, R. and Trias, A. (1981). *Phys. Rev.* **D23**, 553.

Gambini, R. and Trias, A. (1986). *Nuc. Phys.* **B278**, 436.

Gamow, G. (1970). *My World Line, An Informal Autobiography.* Viking Adult.

Garay, J., Martín-Benito, M., and Mena-Marugán (2010). *Phys. Rev.* **D82**, 044048.

Giesel, K., Hoffmann, S., Thiemann, T., and Winkler, O., (2007). *Class. Quant. Grav.* **27**,055005; 055006.

Giesel, K. and Thiemann, T. (2007). *Class. Quan. Grav.* **24**, 2465; 2499; 2565.

Giles, R. (1981). *Phys. Rev.* **D24**, 2160.

Gleiser, R. and Kozameh (2001). *Class. Quant. Grav.* **20**, 4375.

Goenner, H. (2004). *Liv. Rev. Rel.* **7**, 2.

Groenewold, H. (1946). *Physica* **12**, 405.

Hanson, A., Regge, T., and Teitelboim, C. (1976). "Constrained Hamiltonian Systems." Accademia Nazionale Lincei, Rome. https://scholarworks.iu.edu/dspace/handle/2022/3108

Hartle, J. (2003). *Gravity: An Introduction to Einstein's General Relativity.* Benjamin Cummings, New York.

Hawking, S. (1971). *Phys. Rev. Lett.* **26**, 1344.

Helling, R. and Policastro, G. (2004). "String quantization: Fock vs. LQG representations," arXiv:hep-th/0409182.

Henneaux, M. and Teitelboim, C. (1992). *Quantization of Gauge Systems.* Princeton University Press, Princeton.

Huang, K. (1992). *Quarks, Leptons and Gauge Fields.* World Scientific, Singapore.

Israel, W. (1967). *Phys. Rev.* **164**, 1776.

Jacobson, T. (1995). *Phys. Rev. Lett.* **75**, 1260.

Jacobson, T. and Smolin, L. (1988). *Nucl. Phys.* **B299**, 295.

Károlyházy F., Frenkel A., Lukács, B. (1986). "Gravity in the reduction of the wavefunction," in R. Penrose and C. Isham, (eds.), *Quantum Concepts in Space and Time.* Oxford University Press, Oxford.

Kaul, R. and Majumdar, P. (2000). *Phys. Rev. Lett.* **84**, 5255.

Kiefer, C. (1988). *Phys. Rev.* **D38**, 1761.

Kiefer, C. (2006). *Quantum Gravity.* Oxford Science Publications, Oxford.

Krasnov, K. (1997). *Phys. Rev.* **D55**, 3505.

Kruskal, M. (1960). *Phys. Rev.* **119**, 1743.

Kogut, J. and Susskind, L. (1975). *Phys. Rev.* D11, 395.

Kuchař, K. (1992). "Time and Interpretations of Quantum Gravity," in G. Kunstatter, D. Vincent and J. Williams (eds.), *Proceedings of the 4th Canadian Conference on General Relativity and Relativistic Astrophysics.* World Scientific, Singapore. To Be reprinted in Int. J. Mod. Phys. D

Laddha, A. and Varadarajan, M. (2010). "The Hamiltonian constraint in Polymer Parametrized Field Theory," *Phys. Rev.* D83:025019.

Lauscher, O. and Reuter (2002). *Phys. Rev.* **D66**, 025026; *Int. J. Mod. Phys.* **A17**, 993–1002; *Class. Quant. Grav.* **19**, 483–492; *Phys. Rev.* **D65**, 025013.

Lauscher, O. and Reuter, M., (2005). *JHEP* 0510, 050.

Lewandowski, J., Okołów, A., Sahlmann, H., and Thiemann, T. (2006). *Commun. Math. Phys.* **267**, 703.

Lloyd, S. and Ng, Y. (2004). *Scientific American*, November issue, 56.

Mattingly, J. (2006). *Phys. Rev.* **D73**, 064025.

Meissner, K. (2004). *Class. Quant. Grav.* **21**, 5245.

Mitrofanov, I. (2003). *Nature*, **426**, 139.

Ng, Y. and van Dam, H. (1995). *Annals N. Y. Acad. Sci.* **755**, 579.

Nicolai, H., Peeters, K., and Zamaklar, M. (2005). *Class. Quant. Grav.* **22**, R193.

Nicolai, H. and Peeters K. (2007). *Lect. Notes Phys.* **721**, 151.

Padmanabhan, T. (2010). *Rept. Prog. Phys.* **73**, 046901.

Page, D. N. and Geilker, C. D. (1981). *Phys. Rev. Lett.* **47**, 979.

Page, D. and Wootters, W. (1983). *Phys. Rev.* D27, 2885.

Penrose, R. (1971). in T. Bastin, (ed.), *Quantum Theory and Beyond.* Cambridge University Presss, Cambdridge.

Percacci, R. (2006). *Phys. Rev.* **D73**, 041501.

Perez, A. (2004). "Introduction to loop quantum gravity and spin foams," [arXiv:gr-qc/0409061][accessed 11 March 2011].

Perez, A. (2006). *Phys. Rev.* **D73**, 044007.

Perez, A. (2011). "The spin foam approach to quantum gravity." In preparation for *Liv. Rev. in Rel.* and "Recent advances in spin foam models" in preparation for *Pap. in Phys.*

Peskin, M. and Schroeder, D. (1995). *An Introduction to Quantum Field Theory*. Addison Wesley, New York.

Ponzano, G, Regge, T. (1968). "Semiclassical limit of Racah coefficients" in F. Bloch (ed.), *Spectroscopic and Group Theoretical Methods in Physics*. North-Holland Publ. Co., Amsterdam.

Regge, T. (1961). *Nuo. Cim.* XIX, 559.

Requardt, M. (2008). "About the minimal resolution of space-time grains in experimental quantum gravity", arXiv:0807.3619 [gr-qc][accessed 11 March 2011].

Reisenberger, M. and Rovelli, C. (1997). *Phys. Rev.* **D56**, 3490.

Rindler, W. (1977). *Essential Relativity*. Springer-Verlag, New York.

Romano, J. (1993). *Gen. Rel. Grav.* **25**, 759.

Rovelli, C. (1991). *Phys. Rev.* **D43**, 442.

Rovelli, C. (1996). *Phys. Rev. Lett.* **77**, 3288.

Rovelli, C. (2002). "Notes for a brief history of quantum gravity," in V. Gurzadyan, R. Jantzen, R. Ruffini (eds.), *Proceedings of the 9th Marcel Grossmann Meeting*. World Scientific, Singapore.

Rovelli, C. (2007). *Quantum Gravity*. Cambridge University Press, Cambridge.

Rovelli, C. and Smolin, L. (1988). *Phys. Rev. Lett.* **61**, 1155.

Rovelli, C. and Smolin, L. (1990). *Nucl. Phys.* **B133**, 80.

Rovelli, C. and Smolin, L. (1995). *Phys. Rev.* **D52**, 5743.

Rovelli, C. and Speziale, S. (2003). *Phys. Rev.* **D67**, 064019.

Rovelli, C. and Upadhya, P. (1998). "Loop quantum gravity and quanta of space: A primer," arXiv:gr-qc/9806079[accessed 11 March 2011].

Sahlmann, H. and Thiemann, T. (2006). *Class. Quan. Grav.* **23**, 867; 909.

Salecker, H. and Wigner, E. (1958). *Phys. Rev.* **109**, 571.

Singh, P. (2011). "A pedestrian guide to loop quantum cosmology" in preparation for *Pap. in Phys.*

Schlamminger, S., Choi, K.-Y., Wagner, J., Gundlach, J., and Adelberger, E. (2008). *Phys. Rev. Lett.* **100**, 041101.

Schutz, B. (2009). *A First Course in General Relativity*. Cambridge University Press, Cambridge.

Smolin, L. (2002). *Three Roads to Quantum Gravity*. Basic Books, New York.

Stelle, K. (1977). *Phys. Rev.* **D16**, 953.

Szekeres, G (1960). *Pub. Mat. Debrecen* 7, 285. Reprinted in (2002) *Gen. Rel. Grav.* **34**, 2001.

Thiemann, T. (1996). *Phys. Lett.* **B380**, 257.

Thiemann, T. (1998). *J. Math. Phys.* **39**, 1236.

Thiemann, T. (2006). *Class. Quan. Grav.* **23**, 2249.

Thiemann, T. (2007). *Lect. Notes Phys.* **721**, 185.

Thiemann, T. (2008). *Modern Canonical Quantum General Relativity.* Cambridge University Press, Cambridge.

van Hove, L. (1951). *Proc. Roy. Acad. Sci. Belg.*, **26**, 1.

Weinberg, S. (1979). "Ultraviolet divergences in quantum theories of gravitation," in S. Hawking and W. Israel (eds.), *General Relativity: An Einstein Centenary Survey.* Cambridge University Press, Cambridge.

Will, C. M. (2005). *Living Rev. Rel.* **9**, 3.

Woodard, R. (2009). "How far are we from the quantum theory of gravity?," *Rept. Prog. Phys.* 72:126002, 200.

Zwiebach, B. (2009). *A First Course on String Theory.* Cambridge University Press, Cambridge.

Index